Undergraduate Texts in Mathematics

Editors

S. Axler
F.W. Gehring
K.A. Ribet

Undergraduate Texts in Mathematics

Abbott: Understanding Analysis.

Anglin: Mathematics: A Concise History and Philosophy.
Readings in Mathematics.

Anglin/Lambek: The Heritage of Thales.
Readings in Mathematics.

Apostol: Introduction to Analytic Number Theory. Second edition.

Armstrong: Basic Topology.

Armstrong: Groups and Symmetry.

Axler: Linear Algebra Done Right. Second edition.

Beardon: Limits: A New Approach to Real Analysis.

Bak/Newman: Complex Analysis. Second edition.

Banchoff/Wermer: Linear Algebra Through Geometry. Second edition.

Berberian: A First Course in Real Analysis.

Bix: Conics and Cubics: A Concrete Introduction to Algebraic Curves.

Brémaud: An Introduction to Probabilistic Modeling.

Bressoud: Factorization and Primality Testing.

Bressoud: Second Year Calculus.
Readings in Mathematics.

Brickman: Mathematical Introduction to Linear Programming and Game Theory.

Browder: Mathematical Analysis: An Introduction.

Buchmann: Introduction to Cryptography.

Buskes/van Rooij: Topological Spaces: From Distance to Neighborhood.

Callahan: The Geometry of Spacetime: An Introduction to Special and General Relativity.

Carter/van Brunt: The Lebesgue–Stieltjes Integral: A Practical Introduction.

Cederberg: A Course in Modern Geometries. Second edition.

Chambert-Loir: A Field Guide to Algebra

Childs: A Concrete Introduction to Higher Algebra. Second edition.

Chung/AitSahlia: Elementary Probability Theory: With Stochastic Processes and an Introduction to Mathematical Finance. Fourth edition.

Cox/Little/O'Shea: Ideals, Varieties, and Algorithms. Second edition.

Croom: Basic Concepts of Algebraic Topology.

Curtis: Linear Algebra: An Introductory Approach. Fourth edition.

Daepp/Gorkin: Reading, Writing, and Proving: A Closer Look at Mathematics.

Devlin: The Joy of Sets: Fundamentals of Contemporary Set Theory. Second edition.

Dixmier: General Topology.

Driver: Why Math?

Ebbinghaus/Flum/Thomas: Mathematical Logic. Second edition.

Edgar: Measure, Topology, and Fractal Geometry.

Elaydi: An Introduction to Difference Equations. Second edition.

Erdős/Surányi: Topics in the Theory of Numbers.

Estep: Practical Analysis in One Variable.

Exner: An Accompaniment to Higher Mathematics.

Exner: Inside Calculus.

Fine/Rosenberger: The Fundamental Theory of Algebra.

Fischer: Intermediate Real Analysis.

Flanigan/Kazdan: Calculus Two: Linear and Nonlinear Functions. Second edition.

Fleming: Functions of Several Variables. Second edition.

Foulds: Combinatorial Optimization for Undergraduates.

Foulds: Optimization Techniques: An Introduction.

Franklin: Methods of Mathematical Economics.

(continued after index)

Paul Erdős
János Surányi

Topics in the Theory of Numbers

Translated by Barry Guiduli

With 32 Illustrations

 Springer

Paul Erdős
(deceased)

János Surányi
Department of Algebra and
  Number Theory
Eötvös Loránd University
Pázmány Péter Sétany 1/C
Budapest, H-1117
Hungary
suranyi@cs.elte.hu

Editorial Board

S. Axler
Mathematics Department
San Francisco State
  University
San Francisco, CA 94132
USA
axler@sfsu.edu

F.W. Gehring
Mathematics Department
East Hall
University of Michigan
Ann Arbor, MI 48109
USA
fgehring@math.lsa.umich.edu

K.A. Ribet
Mathematics Department
University of California,
  Berkeley
Berkeley, CA 94720-3840
USA
ribet@math.berkeley.edu

Mathematics Subject Classification (2000): 51-01

Library of Congress Cataloging-in-Publication Data
Erdős, Paul, 1913–
    [Válogatott fejezetek a számelméletbol. English]
    Topics in the theory of numbers / Paul Erdős, János Surányi.
      p. cm. — (Undergraduate texts in mathematics)
    Includes bibliographical references and index.
    ISBN 978-1-4612-6545-0   ISBN 978-1-4613-0015-1 (eBook)
    DOI 10.1007/978-1-4613-0015-1
      1. Number theory.   I. Surányi János, 1918–   II. Title.   III. Series.
    QA241 .E7613 2002
    512′.7–dc21   2002067532

ISBN 978-1-4612-6545-0   Printed on acid-free paper.

© 2003 Springer Science+Business Media New York
Originally published by Springer Science+Business Media, Inc. in 2003
Softcover reprint of the hardcover 1st edition 2003
All rights reserved. This work may not be translated or copied in whole or in part without the written permission of the publisher Springer Science+Business Media, LLC except for brief excerpts in connection with reviews or scholarly analysis. Use in connection with any form of information storage and retrieval, electronic adaptation, computer software, or by similar or dis- similar methodology now known or hereafter developed is forbidden. The use in this publication of trade names, trademarks, service marks and similar terms, even if they are not identified as such, is not to be taken as an expression of opinion as to whether or not they are subject to proprietary rights.

(MVY)

9 8 7 6 5 4 3 2

springeronline.com

Preface to the Second Edition[1]

The first edition of our book has long since been out of print, and many people who regard this book as the source of their first mathematical inspiration, and others whose only contact with the book was in a library (the book has disappeared from many), have been pushing for a long time for a new edition. We happily comply with all these urgings. We would especially like to thank MIKLÓS SIMONOVITS, who not only encouraged our work unwaveringly, but also gave precious help.

The goals of the book as expressed in the preface to the first edition are unchanged. Some sections, however, have undergone significant changes. A new Chapter 2 was written about congruences. The several important theorems appearing in this new chapter originally appeared in the first edition stated in terms of divisibility. However, we felt it useful to introduce the notion of congruence. We have included a more thorough discussion of the subject, including second-degree congruences. Here as well we have chosen an interesting, albeit lesser known, path to follow.

In the nearly 40 long years that have passed since the first edition, many important changes have occurred in all of mathematics, as well as in number theory in particular, and to a certain degree the authors' interests have changed as well. These changes have resulted in two important changes in the book. We have omitted a proof of a number-theoretic theorem about polynomials, as well as the accompanying section on polynomial arithmetic. Furthermore, the section concerning the number of elements in a series no three of whose elements forms an arithmetic progression has been significantly shortened; since the first edition, SZEMERÉDI has established a very deep result for these series. On the other hand, we have included a few new results on so called Sidon sets, sets of numbers that do not contain the difference of any two elements. (See Section 6.21 and subsequent sections.)

We remarked in the preface to the first edition that one charm of the integers is that easily stated problems, which often sound simple, are often very difficult and sometimes even hopeless given the state of our current knowledge. For instance, in a 1912 lecture at an international mathematical congress, EDMUND LANDAU mentioned four old conjectures that appeared hopeless at that time:

---

[1] This is an abridged translation of the original Hungarian text.

- Every even number greater than two is the sum of two primes.
- Between any consecutive squares there is a prime number.
- There are infinitely many twin primes (these are primes that are consecutive odd integers).
- There are infinitely many primes of the form $n^2 + 1$.

Today, his statement could be reiterated. During the more than 80 years that have passed, much intensive research has been conducted on all of these conjectures, and we now know that every large enough odd number is the sum of three primes, and every even number can be written as the sum of a prime and a number that is divisible by at most two different primes; there are infinitely many consecutive odd integers for which one is a prime and the other has at most two prime divisors; for infinitely many n, there is a prime between n^2 and $n^2 + n^{1.1}$; for infinitely many n, $n^2 + 1$ has at most two distinct prime divisors. Unfortunately, the methods used to achieve these rather deep results cannot be generalized to prove the more general conjectures, and given the state of mathematics today, the resolution of these conjectures can still be called hopeless.

As we mentioned, we restricted ourselves to results that can be established by using only elementary tools. This does not mean, however, that their proofs are simple. Juxtaposed with rather simple ideas, some rather deep proofs occur, especially from Chapter 5 onward. Among these, the most difficult is Theorem 11 in Chapter 5 (Sections 20–22.), Theorems 4 and 5 in Chapter 8 (Sections 11*–13* and 16*–18*, respectively). We have marked these sections with stars; for less experienced readers, we recommend skimming these sections during the first reading.

We express our deepest gratitude to IMRE Z. RUZSA for reading through the entire manuscript and to IMRE BÁRÁNY for reading through Chapter 4; they greatly helped our work. It would be very difficult to list all those people whose comments and suggestions from the first edition have helped improve this edition. Here we collectively express our appreciation to one and all. For selfless help with the revisions of the book, we express our deepest thanks to ANTAL BALOG, RÓBERT FREUD, KÁLMÁN GYŐRY, JÁNOS PINTZ, ANDRÁS SÁRKÖZY, and VERA T. SÓS and additionally to all those people who supported us with their comments and advice.

Budapest, August 1995   Paul Erdős, János Surányi

Preface to the First Edition[2]

The numbers we know best are the integers, but these are perhaps the most elusive as well. Number theory, the branch of mathematics that studies their properties, is a repository of interesting and quite varied problems, sometimes impossibly difficult. We have gathered together a collection of problems from various topics in this field of mathematics that we find beautiful, intriguing, and from a certain point of view instructive. We hope that others take pleasure in them as well.

In addition to reveling in the beauty of the problems themselves, we have tried to give glimpses into the deeper related mathematics. We endeavored to show the living mathematics, giving examples of problems that can be solved using elementary tools, and which often have related problems whose solutions require very difficult lemmas and deep ideas; in fact, among these related problems we often come across ones whose solutions seem hopeless in light of the present state of knowledge.

In our book we present only problems whose solutions can be obtained using elementary methods. We do not assume any prior knowledge of number theory.

We tried to use only a few results from other areas of mathematics, which we were obliged to use without proof because their proofs fall outside the scope of this book, and by trying to include their proofs, the book would have grown too big. Among these, we list the most important ones. The reader can find their proofs in any elementary textbook, but the discussions within this book can be followed without any difficulty if the reader accepts the results without proof.

For many of the proofs we try to provide motivation as to why we approach problems in the given manner, and we try to present the important lines of thought needed to arrive at the solution. In these instances, we have borrowed wisdom from GYÖRGY PÓLYA's books, and we would like to thank professors RÓZSA PÉTER and LÁSZLÓ KALMÁR. (It is another question as to how well we have succeeded.)

We have included exercises after the different problem topics. In some instances these are easy problems whose solutions build upon or are based on the established results. The other instances the problems give results that

---

[2] This is an abridged translation of the original Hungarian text.

extend the themes discussed and are of varying degrees of difficulty. The most difficult of these we have indicated by a star. We give hints to their solutions in the appendix of this book. When the problems are due to other authors, we have indicated their sources, in so much as it was possible to determine.

We have divided the chapters into sections. In the interest of readability, however, we have not given these names. The reader will find the detailed content in the table of contents, with the sections grouped by subject. The sections, theorems, formulas, and footnotes are numbered in increasing order, starting anew in each chapter. We refer to these using only the number when they are in the same chapter, and using the number prefixed by the chapter number when they are in other chapters.

The chapters of the book build little upon each other, except for the fundamental notions given in the first chapter, and can essentially be read independently of each other after the first chapter, with only the material of the third and fourth chapters [with the new numbering] being related.

PÁL TURÁN strongly supported our work, starting from the selection of subject matter, and furthermore we received many interesting comments from him and RÓZSA PÉTER upon reading our entire manuscript. With gratitude, we thank them both.

Budapest, July 1959   Paul Erdős, János Surányi

Preface to the English Translation

The death of Professor PAUL ERDŐS on September 20, 1996, in Warsaw, was a great loss to the world mathematical community. I can attest to the fact that until the very end he kept a watchful eye on this translation and proposed several new results that arose since the writing of the second Hungarian edition. The evening before he left for Warsaw we were together in my apartment discussing the translation.

For the translation of the manuscript we express our deepest gratitude to the translator, BARRY GUIDULI, who not only translated the text but also provided many interesting comments and suggestions. Best thanks are also due to ROMY VARGA, who assisted him greatly in the translation.

We also thank ZOLTÁN KIRÁLY for indispensable technical help. To the Alfréd Rényi Mathematical Research Institute of the Hungarian Academy of Science, as well as to the Computer Science and Mathematics Departments of Eötvös Loránd University, we are indebted for their continuous support. Without their gracious help and support, this project would not have been realized.

Budapest, April 2000 — János Surányi

Contents

Note: Chapters are divided into sections, and groups of sections form unified topics; this is indicated in the table of contents by indentation. Thus, for example, in Chapter 1, Sections 1 through 3 form a topic, as do Sections 4 and 5.

Facts Used Without Proof in the Book

1. By $n!$ we mean the product of the first n positive integers, and we read it n *factorial*. This is also the number of distinct orderings (permutations) of a set of n elements. We also use $0!$, whose value we define to be 1.
2. The nth power of a two-element sum has the polynomial representation

$$(a+b)^n = \binom{n}{0}a^n + \binom{n}{1}a^{n-1}b + \cdots + \binom{n}{i}a^i b^{n-i} + \cdots + \binom{n}{n}b^n,$$

where the coefficients are

$$\binom{n}{0} = 1, \quad \binom{n}{i} = \frac{n(n-1)\cdots(n-i+1)}{i!} = \frac{n!}{i!(n-i)!} \quad \text{if} \quad 1 \le i \le n,$$

which are integers. The symbol is read "n choose i," and these are called *binomial coefficients*. We can choose i-element subsets from an n-element set in $\binom{n}{i}$ ways (where we are not concerned with the order of the i elements). This shows that these must be integers, which is not clear from the formulas given above.
3. More generally, the nth power of the sum $a_1 + a_2 + \cdots + a_k$ is the sum of all terms of the form

$$\frac{n!}{i_1! i_2! \cdots i_k!} a_1^{i_1} a_2^{i_2} \cdots a_k^{i_k},$$

where $i_1, i_2, \ldots, i_k$ range over all ordered k-tuples of nonnegative integers whose sum is n.
4. The infinite series $1 + z + z^2 + \cdots$ converges for all complex numbers z for which $|z| < 1$, and the sum is $1/(1-z)$.
5. By log we mean the natural logarithm, whose base is the number e. The following inequality holds:

$$\log(1+x) \le x,$$

for all $x > -1$, or in the exponential form, for every real number x,

$$e^x \ge 1 + x.$$

Sometimes we may write $\exp(x)$ in place of e^x.

6. The following two inequalities hold for the sum of the reciprocals of the first n positive integers:

$$1 + \log n \geq 1 + \frac{1}{2} + \cdots + \frac{1}{n} > \log(n+1).$$

7. The function $\log x$ grows more slowly than x to any positive power, which means that for any positive δ and any positive h, there exists a number K (dependent on δ and h) such that

$$\frac{\log x}{x^\delta} < h$$

holds for all $x > K$.

8. The greatest integer not larger than a real number x is called the *integer part*[3] *of* x and is written $[x]$. This is the unique integer for which the following relation holds:

$$[x] \leq x < [x] + 1.$$

9. The difference $x - [x]$ is called the *fractional part of* x and is written $\{x\}$. (In general, this is not a rational number, i.e., it cannot be expressed as the ratio of two integers.)

10. a polynomial with complex or rational coefficients cannot be written as the product of two polynomials of smaller degree with the same type of coefficients, then the roots are simple. (Such polynomials are called *irreducible.*)

These formulas will be used from time to time in the text, occasionally without reference. If we refer to the ith point, we will write, "Fact i."

---

[3] This is also called the *floor of* x, and the smallest integer not less than x is called the *upper integral part of* x, or the *ceiling of* x. These are written $\lfloor x \rfloor$ and $\lceil x \rceil$, respectively.

1. Divisibility, the Fundamental Theorem of Number Theory

1. Counting and the numbers that thus came forth are among the earliest achievements of mankind's awakening intellect. As numbers came to being, their intriguing properties were revealed, and symbolic meanings were assigned to them. Besides portending fortune or doom, numbers afforded a mathematical expression to many other aspects of existence. For instance, the ancient Greeks considered the divisors of a number that are less than the number itself to be its parts; indeed, they so named them. And those numbers that rise up from their parts, like Phoenix, the bird that according to legend rises up from its own ashes, were viewed as the embodiment of perfection. Six is such a *perfect* number, since it is the sum of its parts 1, 2, and 3; 28 and 496 are also perfect. EUCLID (third century B.C.) already knew that a number of the form $2^{n-1}(2^n - 1)$ is perfect if the second factor is prime.[1] It was LEONHARD EULER (1707–1783), more than two thousand years later, who first showed that any even perfect number must be of this form. To this day it is still unknown whether or not there exist odd perfect numbers.

PYTHAGORAS (sixth century B.C.) is said to have remarked that two of his students were true friends, like 220 and 284. These numbers were called *amicable*, since each is the sum of the parts of the other.

These were the only amicable numbers known until the modern renaissance of mathematics. At that time, mathematicians of the seventeenth and eighteenth centuries found them in abundance. They even found pairs of odd numbers, such as

$$12\,285 = 3^3 \cdot 5 \cdot 7 \cdot 13 \quad \text{and} \quad 14\,595 = 3 \cdot 5 \cdot 7 \cdot 139.$$

Curiously, however, the smallest pair after the one known from antiquity, 1184 and 1210, was found only in 1766 by a 16-year-old Italian student by the name of NICCOLÒ PAGANINI.[2] Despite the multitude of amicable pairs of numbers that have been discovered, we know little about them in general.

---

[1] EUCLID did not have this notation to use. He described it in the following way: if the sum of a geometric sequence starting with 1 and having a ratio of 2 is prime, then multiplying this sum by the last element of the sequence yields a perfect number. (Euclid: Elements. SIR THOMAS L. HEATH, New York, Dover Publications, 1956.)

[2] Not to be confused with the famous violin virtuoso of the same name who lived much later.

Among other things it is not known whether an even number can be paired with an odd number.

One could add many more examples, but perhaps these few suffice to hint that, familiar as the whole numbers may be, they hold many secrets. Searching for their secrets promises to be an interesting task, and often not an easy one.

2. The sum and product of two integers is an integer, as is their difference; the quotient of two integers is, however, seldom an integer, and it is not immediately clear for which pairs of integers this property holds. We can, however, perform division with remainder so that the quotients are always integers, as long as the divisor is not zero. We define this in the following way:

Theorem 1 (Remainder Theorem) *For all numbers a and b, b ≠ 0, there is an integer c and a number d such that*

$$a = bc + d \quad and \quad 0 \le d < |b|,$$

and only one such c and d exist.

We say that a divided by b has *quotient* c with *remainder* d. The validity of the theorem is self-evident if we look at the number line. We mark all multiples of b on the number line, getting a series of equidistant points (see Figure 1). If a happens to lie on one of these numbers, say bc, then this is our c, and we let $d = 0$. If a falls between two multiples, we take the c corresponding to the multiple of b just before a and then d is the distance from bc to a. These choices satisfy the conditions of the theorem; if we were to take some other c and d satisfying $a = bc + d$, then this d would be either negative or greater than $|b|$. With this observation we have verified the uniqueness of c and d.

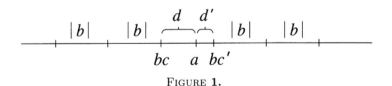

FIGURE 1.

In the case where a is closer to the multiple of b to its right, it makes more sense to choose c corresponding to this larger multiple. This leads us to the following theorem:

Theorem 1′ *For all numbers a and b, where b ≠ 0, there is an integer c′ and a number d′ such that*

$$a = bc' + d' \quad and \quad -\frac{|b|}{2} < d' \le \frac{|b|}{2},$$

and only one such c' and d' exist.

We leave the proof to the reader. If we define the remainder upon division from Theorem 1, then we mean the smallest possible nonnegative remainder. If we define it by Theorem 1', then we mean the remainder with smallest possible absolute value.

The reader should note that in the above theorems we require only that the quotients be integers; the other numbers that occur are arbitrary. In this general form we may use the theorem for the measurement of angles, for instance by radians. Given any such angle α, we may characterize it by an angle δ between 0 and 2π or an angle δ' between $-\pi$ and π (depending on which is more convenient for us to use), where

$$\alpha = 2\pi c + \delta, \qquad 0 \le \delta < 2\pi,$$

and

$$\alpha = 2\pi c' + \delta', \qquad -\pi < \delta' \le \pi,$$

where c and c' are integers by Theorems 1 and 1', respectively.

Exercises:

1. Prove Theorem 1'.

In the following problems we consider positive integers written in base 10 notation unless otherwise specified.

2. Find rules for divisibility for numbers where the divisor is 2, 4, 8, 5, or 25.

3. Prove that a number and the sum of its digits have the same remainder upon division by 9. (In particular, this says that a number is divisible by 9 if and only if the sum of its digits is divisible by 9.) What divisibility rule follows from this for division by 3?

4. Given an integer, consider the difference between the sum of its digits in odd positions (counting from the right) and those in even positions. Show that this difference has the same remainder as the number itself when divided by 11. What divisibility rule does this give us?

5. Partition the digits of a number into groups of two and three, respectively, starting from the right. Treat these as two- and three-digit numbers, respectively, and proceed similarly as in Exercises 3 and 4. What divisibility rules arise?

6. (a) Remove the last digit from a number and subtract twice this digit from the new (shorter) number. Show that the original number is divisible by 7 if and only if this difference is divisible by 7. (By repeating this procedure, one obtains a rather useful divisibility rule.)

(b) What if we add two times the digit to the new number instead of subtracting it?

(c) Give similar rules for divisibility by 13 and 17. (Use the fact that $3 \cdot 13 = 39$ and $3 \cdot 17 = 51$.)

7. (a) Prove that if n is an integer larger than one, then every natural number can be written in *base* n. This means that every natural number can be written as:

$$\sum_{0 \le j \le k} c_j n^{k-j}, \quad 0 \le c_j \le n-1, \quad j = 0, 1, \ldots, k,$$

for some k, and disregarding leading zeros, this presentation is unique.

(b) More generally, if $n_1, n_2, \ldots, n_j, \ldots$ is a list of integers greater than 1, then every positive integer can be written in the following *varying base representation*:

$$c_0 + \sum_{1 \le i \le k} c_i \prod_{1 \le j \le i} n_j, \quad 0 \le c_i \le n_{i+1} - 1, \quad i = 0, 1, \ldots, k,$$

for some k. Disregarding leading zeros, this too is unique. (If $n_j = n$ for all j, then this is just part (a). What sort of representation arises by setting $n_j = j + 1$?)

8. Give divisibility rules for the following base systems: 6, 12, 5.

9. (a) If n is an integer greater than 1, expand the following polynomial:

$$F(x) = \left(1 + x + \cdots + x^{n-1}\right)\left(1 + x^n + \cdots + x^{(n-1)n}\right) \cdots$$
$$\cdots \left(1 + x^{n^k} + \cdots + x^{(n-1)n^k}\right),$$

and use the result to prove the claim in Exercise 7 Part (a).

(b) Give a similar proof for Exercise 7 Part (b).

10. Given an integer $n > 2$, prove that any integer can be represented in the following form:

$$\sum_{0 \le j \le k} d_j n^j, \quad -\frac{n}{2} < d_j \le \frac{n}{2}, \quad j = 0, 1, \ldots, k.$$

11. There is a plate of cherries on a table. In our absence, someone serves the cherries in the following way: For every 5 cherries taken from the plate, one is put on a second plate and the other four are put in a serving bowl. This continues until there are fewer than five left on the original plate. From the second plate the procedure continues, using a third plate until there are fewer than five cherries left on the second plate. This continues until there are fewer than five cherries on the

last plate. The bowl is then put away. We are presented with only the plates and asked to determine how many cherries there were at the beginning. How do we do this? How many cherries were there originally if on 4 plates we now have 2, 4, 0, and 3 cherries, in this order? How would we do it if instead of groups of five, the cherries were served in groups of two or groups of ten (one cherry placed on the new plate, the remaining placed in the bowl)?

12. We can multiply two numbers together in the following way. Write the two numbers down next to each other. Divide the first in half, rounding down to an integer, and write the result below it. Double the second number, writing this result below it. We continue this halving/doubling until we are left with 1 in the first column. Cross out all those numbers in the second column that are opposite an even number and add the remaining numbers in this column together to get the product. Prove that this works. The adjacent example illustrates how to multiply 73 by 217 using this method.

73	217
36	~~434~~
18	~~868~~
9	1 736
4	~~3 472~~
2	~~6 944~~
1	13 888
	15 841

3. Measuring intervals of equal length on a number line provides a quite simple sequence of points. It is rather more difficult to see what happens if we measure off arcs of equal length on a circle, but here we can also discover some simple properties. To start with, we will consider only sequences of arcs where the length is such that we eventually meet up with an earlier point. Later, in Chapter 3, we will return to the case where the sequence does not meet up with itself.

Consider a circle with perimeter c, and a point P_0 on it. From P_0, measure off arcs of equal length in one direction. Call this direction *positive* and label the points P_1, P_2, etc. (see Figure 2). We are assuming that the sequence eventually runs into itself. Let P_t be the first point that coincides with another point P_j of the sequence. Then j must be 0, for if this were not true, then P_{t-1} would have already fallen on P_{j-1}.

Continuing to measure off intervals, we land on points $P_1, P_2, \ldots, P_{t-1}$, and with P_{2t} we land again on P_0. We see that all points falling on P_0 have an index that is a multiple of t, and before we mark off such a point, we have landed on every point exactly the same number of times.

We use these observations to prove the following theorem.

Theorem 2 (Four Number Theorem) *If a and c are positive numbers and b and d are positive integers such that*

$$ab = cd, \tag{1}$$

then there exists a positive number r and positive integers s, t, and u such that the following equalities hold:

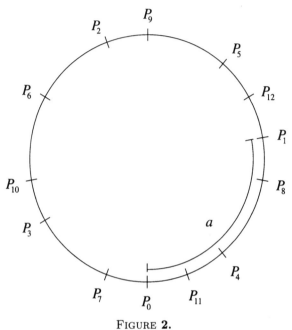

FIGURE **2.**

$$a = rs, \quad b = tu, \quad c = rt, \quad d = su.$$

If, in addition, a and c are integers, then r may be taken to be an integer.

If $a = c$, then $b = d$ and we may choose $r = a$, $s = t = 1$, and $u = b$. If a and c are not equal, we assume that $a < c$ (otherwise, we may change the roles of a and c). We may interpret (1) in the following, geometric, way: Consider a circle of circumference c, and a point P_0 on it. We mark off arcs of length a, labeling the points in sequence P_1, P_2, through P_b. Equation (1) says that P_b coincides with P_0 and that we have gone around the circle a total of d times. If the tth point is the first that coincides with a previous one, then by the earlier observations, this point is P_0. The entire sequence returns to P_0 u times, and lands on each of the t distinct points $P_0, P_1, \ldots, P_{t-1}$ exactly u times, so

$$b = tu.$$

If we start at any point, say P_j, and mark off arcs of length a, we land on P_0 after $b - j$ steps, and continuing, we have $P_1, P_2, \ldots, P_{j-1}, P_j$, getting the entire sequence back. We can look at this in another way: We rotate the circle, putting P_0 in P_j's position, and starting at P_j we get the original sequence. This holds true for all points P_j, and the perimeter of the circle is therefore divided up into arcs of equal length by the t different points. Call this length r, and we have

$$c = rt.$$

Let s be the number of points on the arc between P_0 and P_1 in the positive direction, counting P_1 but not P_0. These points divide the arc into s equal parts of length r, and we have

$$a = rs.$$

Marking off the original sequence, we go around the circle d times. Every time we go around, we hit a point on the arc from P_0 to P_1 (again, not including P_0), hitting each a total of u times. This gives

$$d = su.$$

Thus we have shown the existence of the numbers r, s, t, and u. To verify the last claim, we observe that if the perimeter c and the arc length a are both integers, then every point is an integral number of arc lengths from P_0, and in particular, r is an integer.

Exercises:

   13. Prove that the four number theorem remains true even if we remove the assumption that the numbers are positive.

   14. Prove the four number theorem using induction on $|b|$ instead of geometry.

4. The questions raised in the introduction were related to the divisibility of integers. The four number theorem will play an important role in establishing fundamental relations for division.

It will be advantageous to define the concept of divisibility slightly differently from that of the ancient Greeks; in particular, we will consider the number itself to be one of its divisors. In accordance with this, we say that an integer a *is a divisor of* an integer b if there exists an integer c satisfying the equation

$$b = ac.$$

In this case, we also say that b *is a multiple of* a and that b is *divisible* by a. If a is a divisor of b, we express this symbolically[3] as $a \mid b$; otherwise, we write $a \nmid b$

In the remainder of the book we will concern ourselves principally with integers, and unless we specify otherwise, all numbers should be considered as integers.

The properties of divisibility listed here follow easily from the definition. We will often use them without reference, and for that reason it is useful to see them once presented all together. The proofs are left to the reader. The letters used in the following relations represent arbitrary integers.

---

[3] Unfortunately, a symmetric symbol is used for an asymmetric relation; since this symbol is so widely accepted, we make no attempt to introduce another.

(a) $1 \mid a$ and $-1 \mid a$;

(b) $a \mid ca$, and in particular, $a \mid a$, $a \mid -a$, and $-a \mid a$ $(= -(-a))$;

(c) If $a \mid b$, then $a \mid bc$.

Among the divisors of a, we call 1, -1, a, and $-a$ its *trivial divisors*, and the other positive divisors, those smaller than a, we call its *proper divisors*. The numbers 1 and -1 are distinguished by their role in property (a) as being divisors of very integer. It is customary to call them *units*.

Often we will deal with a set of numbers that do not have a common divisor other than the units. We will call these numbers *relatively prime.*[4]

It is obvious from the definition that if we include a new number in a set of relatively prime numbers, then the resulting numbers are still relatively prime. It is, however, conceivable that if we have more than two numbers that are relatively prime, some of them may have a common divisor other than a unit. For example, the numbers 20, 36, and 45 are relatively prime, yet any two share a common divisor greater than one. Occasionally, we will be interested in numbers that are pairwise relatively prime.

The number 0 is also special in terms of divisibility.

(d) $a \mid 0$;

(e) 0 is only a divisor of zero.[5]

(f) If $a \mid b$ and $b \mid c$, then $a \mid c$.

Exercise:

15. Show that if two numbers are relatively prime, then any divisor of the first and any divisor of the second are also relatively prime.

5. The relationship in property (b) is *reflexivity* of division; that in property (f) is *transitivity*. Division, in general, is not a symmetric property. We can make this statement more precise in the following way:

(g) If $a \mid b$ and $b \mid a$, then $|a| = |b|$.

The result is not significantly weaker than if the two numbers were equal, since a number and its opposite have the same multiples and the same divisors. In symbolic notation, this is

(h) If $a \mid b$, then $-a \mid b$ and $a \mid -b$.

---

[4] We will not introduce a symbol for being relatively prime, since this can easily be expressed by use of the largest common divisor, soon to be introduced. We emphasize here, however, that this definition is independent of a greatest common divisor, which in some cases does not even exist.

[5] This property seemingly contradicts the fact that it is not possible to divide by 0. For division we require that the quotient be properly defined, but for divisibility the existence of at least one c such that (in the above case) $0 = c0$ is enough.

From the point of view of divisibility, a number and its opposite do not differ significantly. In keeping with custom, we call them *associates*[6] of each other. We may also say that the theorems relating (only) to divisibility actually apply to the classes of associated numbers. This means that it is enough to consider divisibility questions by taking one representative from each class, for example the nonnegative one, and this is often what will be done.

We should still examine the connection of divisibility with both addition and subtraction. (And with these properties, the last remark becomes obsolete.)

(i) If $a \mid b$ and $a \mid c$, then $a \mid b + c$.

This is obviously true for sums with arbitrarily many terms. Even the following more general property is true:

(j) If b_i are integers such that $a \mid b_i$, and c_i are arbitrary integers ($i = 1, 2, \ldots, k$), then $a \mid \sum_{i=1}^{k} b_i c_i$.

Often we will use the following argument (which follows from the previous property).

(k) With the assumptions made above, d is divisible by a if and only if the sum $\sum_{i=1}^{k} b_i c_i + d$ is divisible by a.

An additional important connection relates to the size of divisors.

(l) If $a \mid b$ and $b \neq 0$, then $|a| \leq |b|$.

From this it follows that

(m) Every number except 0 has finitely many divisors.

We finish with a property that we will use often.

(n) If $a \mid b$, then $ca \mid cb$, and if $c \neq 0$, then the first relation follows from the second.

Exercises:

16. Prove the divisibility properties listed above.

17. Prove that if we divide the numbers $a_1, a_2, \ldots, a_k$ by their greatest common divisor, then the resulting numbers are relatively prime.

18. Prove that if k and l are positive integers and a^k and b^l are relatively prime, then a and b are relatively prime too. (We note that the converse of this statement is also true, but its proof is considerably harder. See Exercise 25.)

---

[6] The idea of units and associates may seem forced. Later we will extend the definition of divisibility, for example divisibility of polynomials with real or rational coefficients. In these cases, we can have more than one, even infinitely many, units. (In these examples the units are nonzero constants, and polynomials have infinitely many associates.) These concepts will be useful in these cases.

6. We saw that ±1 (the units) and 0 play special roles in terms of divisibility: ±1 have just two trivial divisors, and 0 has three; all other numbers have four trivial divisors. The units do not have other divisors besides the trivial ones, but this property is not unique to the units. It is also shared with many other numbers, for example,

$$2, 3, 5, 7, 11, 73, 137, 139, 3359, 3361, 7963, 10037, 10039,$$

and of course many more. We call such numbers *prime numbers*. Those numbers that can be written as the product of two numbers of smaller absolute value are called *composite numbers*.

Theorem 3 *Every number larger than one has a prime divisor.*

Property (m) states that there are only finitely many divisors, and property (b) guarantees that every divisor's absolute value is also a divisor. We take the smallest divisor among those larger than 1. This is a prime number, for if we could decompose it into the product of two smaller positive numbers, neither of which is 1, then we would have a number larger than one and smaller than our smallest divisor, which is a divisor of this smallest divisor. By property (f) it is also a divisor of our original number, contradicting the fact that the chosen divisor was the smallest.

After the list of primes above we said that there were many more. More exactly, we claim there are infinitely many.

Theorem 4 *There are infinitely many prime numbers.*

We prove the theorem indirectly. Assume that there are only finitely many primes, and that we have written them all down, $p_1, p_2, \ldots, p_k$. We next consider a number that is divisible by all primes, and then alter it a little. For example, we consider the number:

$$p_1 p_2 \cdots p_k + 1.$$

Every listed prime gives a remainder of 1 upon division. By Theorem 3, this number has a prime divisor; but this cannot be one of those listed above, contradicting that we listed all the primes, so there cannot be finitely many primes.

Exercises:

19. **(a)** Prove that every number can be written as $l^2 k$, where l and k are integers and k is not divisible by a square larger than one.[7]

(b) Use the previous observation to find a new proof of Theorem 4.

20*. Prove that in the list of the integers, there are arbitrarily many consecutive composite numbers. (In other words, between two neighboring primes, there occur arbitrarily large gaps.)

---

[7] Numbers of this type are called *square-free.*

7. The primes are the building blocks of the integers in the sense that we can decompose any composite number into a product of primes. It will be convenient to consider a number as the one-factor product consisting of itself. We can now reformulate our statement in the following way:

Theorem 5 *Every number different from 0 and not a unit can be decomposed into the product of finitely many primes.*

Clearly, 2 and −2 are their own decompositions. We continue through the integers, going by size of absolute value. If we arrive at a prime number, this is its own decomposition. If we arrive at a composite number, we split it up into the product of two numbers of smaller absolute value. Since each of these is smaller, we have already determined their decompositions, and the product of these is a decomposition of the new number.

Exercises:

21. (a) Prove that if f and g are polynomials with rational coefficients and g is not the zero polynomial, then there exist polynomials q and r with rational coefficients satisfying the following equation:

$$f = gq + r,$$

where the degree of r is smaller than the degree of g, or r is the zero polynomial. Additionally, prove that there are only one such q and r.

(b) Is the above statement true for polynomials with real coefficients? How about integer coefficients?

We say that a polynomial f with rational coefficients is divisible by a polynomial g if there is a polynomial h with rational coefficients such that $f = gh$. In a similar way we may formulate the notion of divisibility of polynomials with real coefficients and also those with integer coefficients.

22. Go back through Sections 4–6 and rephrase every property of divisibility of integers in terms of polynomials with rational, real, and integer coefficients, respectively. Determine which are true and which are false.

23. (Continuation) In the three cases, determine whether the four number theorem is true or not.

8. The real importance of primes comes from the fact that a prime decomposition is unique (under a certain interpretation of uniqueness, as we will see later on). Before proceeding with this, we will prove two other important properties of divisibility.

In Exercises 2 through 6 we found divisibility rules for every number from 2 to 9, with the only exception of 6, but this is not necessary, since a number is divisible by 6 if it is divisible by both 2 and 3. On the other hand, it is not true that if a number is divisible by both 6 and 3, then it is divisible by 18; 6 itself is an example of this fact.

It is a common mistake to believe property (c) in the opposite direction (this property says that if $a \mid b$, then $a \mid bc$), i.e., that if a number divides a product of numbers, then it divides one of the factors. For example $9 \cdot 10 = 90$ is divisible by 6, whereas neither 9 nor 10 is. The factors of 6, 2 and 3, are split between 9 and 10. However, we can correct this conclusion if we require that one of our factors and the divisor do not have a common divisor—with the exception of units—meaning that they are relatively prime. In this way we arrive at the following theorem:

Theorem 6 *If a number divides the product of two numbers and is relatively prime to one of the factors, then it must divide the other factor.*

The theorem easily follows from the four number theorem. Assume that $c \mid ab$ and c and a are relatively prime. By the definition of divisibility, there is a number d such that
$$ab = cd.$$
So by the four number theorem, there are numbers $r, s, t,$ and u such that
$$a = rs, \qquad b = tu, \qquad c = rt.$$
So r is a common divisor of a and c, which are relatively prime, so r must be 1 or -1, thus $r^2 = 1$. Using this and the third equality, we see that
$$t = r^2 t = rc,$$
so from the second equality, we get that $b = rcu$; hence c divides b, proving the theorem.

This theorem already appeared in EUCLID's *Elements*, and we will call it *Euclid's lemma*.

Exercises:

24. Prove that if a number divides a product of numbers and is relatively prime to all of them except one, then it must divide that one.

25. (a) Prove that if a number is relatively prime to several numbers, then it is relatively prime to their product as well.

   (b) Prove that if two numbers are relatively prime to each other, then arbitrary positive powers of each are still relatively prime to each other.

9. There are certain numbers that have the property that if they divide a product of numbers, then they must divide one of the factors. The units have this property, as does 0, and 2 and 3 as well. We will say that numbers of this type that are different from 0 and the units have the *prime property*.

It is immediately clear that composite numbers do not have the prime property, as we can write a composite number as a product of numbers having smaller absolute value, and by property (1), the original number cannot divide any of these. As the name suggests, we have the following:

Theorem 7 *The prime numbers are precisely those with the prime property.*

We have seen above that only the prime numbers can have the prime-property. The other direction of the claim follows from Theorem 6 (or from the four number theorem, from which this was deduced). If a prime p divides a product ab, then either p divides a, satisfying the claim of the theorem; or if not, then by property (h), its opposite is also not a divisor of a. The only other divisors of p are 1 and -1, and these are the only divisors common to p and a. This means that p and a are relatively prime, and by Theorem 6, p must divide b.

We defined the primes as not being decomposable into factors of smaller absolute value. Theorem 7 gives another important characterization of primes.

10. We have already alluded to the uniqueness of prime decomposition. We formulate it in the following:

Theorem 8 (Fundamental Theorem of Arithmetic) *The prime factorization of a nonzero number that is not a unit is unique up to the order and signs of the factors.*

In particular, we can rephrase this so that if we have two prime factorizations for a given number, then the factors of them can be paired so that corresponding factors have the same absolute value.

The theorem can be easily deduced from the following theorem, which is a generalization of the four number theorem:

Theorem 2′ *If $a_1, a_2, \ldots, a_j$; $b_1, b_2, \ldots, b_k$ are integers such that*

$$a_1 a_2 \cdots a_j = b_1 b_2 \cdots b_k,$$

then there exist integers t_{uv} $(1 \leq u \leq j, 1 \leq v \leq k)$ such that

$$a_u = \prod_{v=1}^{k} t_{uv} \quad (1 \leq u \leq j), \qquad b_v = \prod_{u=1}^{j} t_{uv} \quad (1 \leq v \leq k).$$

We will first show how the fundamental theorem follows from this theorem, and in the next section we will return to the proof of this theorem.

We assume that the two factorizations in the statement of the theorem are prime factorizations. Then the factors a_n are all primes (indecomposable), and considering the equations $a_u = \prod_{v=1}^{k} t_{uv}$, we see that one of the t_{uv}

$(u = 1, \ldots, j)$ is either a_u or $-a_u$, and the others are ± 1. So among all t_{uv}, exactly j have absolute value larger than 1; the rest have absolute value 1.

The same holds if we consider the b_v's. In the product $b_v = \prod_{u=1}^{j} t_{uv}$, one of the terms is b_v or $-b_v$, and all the others have absolute value 1. Hence there are precisely k terms t_{uv} with absolute value greater than 1. It follows that $j = k$, and the terms t_{uv} having absolute value greater than one give a correspondence between the a_u and the b_v, given by

$$|a_u| = |t_{uv}| = |b_v|.$$

This concludes the proof that Theorem 8 follows from Theorem 2'.

11. The proof of the Theorem 2' follows from a two-step induction, the first on j and the second on k. The case $j = k = 2$ is the four number theorem, which we have already proved.

We will first do induction on j. Assume that the theorem is true for $k = 2$ and $j = m$, for some m. We wish to prove the case $j = m + 1$. We have the equation

$$a_1 a_2 \cdots a_m a_{m+1} = b_1 b_2.$$

We consider $a_m a_{m+1}$ to be one factor, and by the inductive assumption, there exist integers t'_{uv} $(u = 1, \ldots, m; \; v = 1, 2)$ such that

$$a_u = t'_{u1} t'_{u2} \qquad (u = 1, \ldots, m-1), \tag{2}$$

$$a_m a_{m+1} = t'_{m1} t'_{m2}, \tag{3}$$

$$b_v = \prod_{u=1}^{m} t'_{uv} \qquad (v = 1, 2). \tag{4}$$

For (3), the four number theorem guarantees the existence of t_{uv} $(u = m, m+1; \; v = 1, 2)$ such that

$$a_u = t_{u1} t_{u2} \qquad (u = m, m+1),$$

$$t'_{mv} = t_{mv} t_{m+1v} \qquad (v = 1, 2).$$

By defining $t_{uv} = t'_{uv}$ for $u = 1, \ldots, m-1; \; v = 1, 2$, we see that the theorem is satisfied in the case $k = 2$, $j = m+1$ by t_{uv} and hence for all j when $k = 2$.

We can now finish the proof by induction on k, keeping j fixed, showing that the theorem holds for all j and k larger than 1. We leave the details of this step to the reader.

12. The history of the fundamental theorem is rather interesting. EUCLID did not mention it in the *Elements*, although Theorem 6 (Euclid's lemma) and Theorem 7 appear in the book. The fundamental theorem follows from either of these, as we shall soon see, and similarly, these both follow from the fundamental theorem. Hence we are dealing with equivalent statements.

Great progress was made in number theory from the mid seventeenth century, but the fundamental theorem was not mentioned, even though consequences of it were used regularly. In EULER's work, the four number theorem and Theorem 2' are used, both without proof. Most likely he came upon these by way of prime decomposition. LEGENDRE published a book on number theory in 1798, but still the fundamental theorem does not appear. Finally, in GAUSS's famous work *Disquisitiones Arithmeticae* (Studies in Arithmetic), published in 1801, the fundamental theorem is stated and proved. He essentially proves Theorem 7 by induction, and the fundamental theorem follows from that.

We should not be surprised that the fundamental theorem does not appear in EUCLID's *Elements*. We mentioned in footnote 1 of Section 1.1 that the Greeks did not have a general algebraic notation to use, and without such a notation, the formulation of the theorem is hopeless. Perhaps the extraordinary respect for EUCLID, and the fact that he did not state the theorem, led mathematicians to take the theorem for granted.

Exercises:

26. Work out the remaining parts of the proof of Theorem 2'.

27. Call the even numbers E-integers. Clearly, sums, differences, and products of these are also E-integers, and of course the usual properties of these operations also hold.

   (a) Define divisibility among E-integers, and determine which of the properties of divisibility hold in this set.

   (b) What are the units (if there are any)? What are the indecomposable elements?

   (c) Is it true that any nonzero E-integer is either indecomposable or can be factored into a product of indecomposable elements?

   (d) Is the corresponding version of the fundamental theorem true for E-integers?

28. **(a)** Does the remainder theorem hold for E-integers?

   (b) We say that an E-integer has the prime property if whenever it divides a product, it follows that it either divides one of the factors or differs from one of the factors in sign. Which of the E-integers have the prime property?

   (c) Does the appropriate version of the Euclid's lemma hold for E-integers? (With the exception of 0, none of the E-integers are divisors of themselves. Here we include the number itself and its additive inverse among the divisors, as we did in part (b).)

Support negative answers to the above questions with counterexamples.

For the understanding of notions related to divisibility and of the fundamental theorem, we need only the operation of multiplication. D. HILBERT

(1862–1943) raised the question of whether or not it is possible to prove the fundamental theorem from the definition and from the properties of multiplication. He showed that it is not possible, giving a counterexample. The following exercises illustrate this example.

29. Call the positive integers of the form $3k + 1$ the H-integers (after HILBERT). Multiplication restricted to the H-integers clearly remains commutative and associative.

    (a) Show that the product of H-integers is an H-integer.

    (b) Define divisibility among H-integers, and examine the properties in Sections 1.5–1.8, determining which have meaning, and of those, which hold true when restricted to H-integers.

30. **(a)** Define the following concepts for the H-integers: unit, indecomposable, and the prime property. Determine which H-integers have those properties.

    (b) Is it possible to write every H-integer as the product of H-indecomposables?

    (c) Is the corresponding version of the fundamental theorem valid for the H-integers?

31. Define what it means to be relatively prime for the H-integers and decide whether or not Euclid's lemma holds for this set.

32. Repeat all of the above exercises for the set of integers of the form $3k + 1$, positive and negative.

33. Solve Exercises 29–31 for integers of the form $8k + 1$, where k is an arbitrary integer.

13. In the prime factorization of a number different from 0, we may gather the primes of equal absolute value together and write them as a power. We may choose to write all the bases of the powers as the positive prime, and may need to include a factor of -1 to account for signs of the original factors. In this way, we can write the number as

$$n = e p_1^{\alpha_1} p_2^{\alpha_2} \cdots p_r^{\alpha_r}.$$

Here e is either 1 or -1, depending on whether n is positive or negative, the p_i are different positive prime numbers, and the α_i are positive integral exponents.

    In the above decomposition it will sometimes be useful to include primes to the power of 0.[8] This prime power decomposition will be called a *canonical decomposition*. Two canonical decompositions can differ in the order of the

---

[8] We could also say that every integer different from 0 associates a nonnegative exponent to every prime. Only finitely many of the exponents can be positive, and according to the fundamental theorem, this association is unique. In the special case of 1 and -1, every prime gets exponent 0.

factors and in the inclusion of factors with exponent 0, but according to the fundamental theorem, they can differ in nothing more.

14. Before we continue, we should examine a few concepts more closely: the divisors of a number, its multiples, the common divisors of more than one number, and their common multiples.

When dealing with questions of divisibility, we have already seen that one need consider only positive divisors, and we will thus restrict ourselves to positive divisors. Furthermore, all numbers will be considered nonnegative integers, unless otherwise specified.

We know that 0 has every number as a divisor, and that 1 has only one divisor (it has two if we consider all integers and include -1). If n is larger than 1, let

$$n = \prod_{i=1}^{r} p_i^{k_i} \tag{5}$$

be a canonical decomposition and let a be a divisor. This means that there is a number b such that

$$n = ab.$$

If we replace the numbers a and b by their canonical decompositions and collect powers with the same base, if there are any, then by the fundamental theorem, we must get the decomposition in (5). This means that in the decompositions of a and b, only those primes can appear that appear in the decomposition of n and their exponents cannot exceed those in the decomposition of n, and therefore the n in (5) can have only the following numbers as its divisors:

$$a = \prod_{i=1}^{r} p_i^{j_i}, \quad \text{where} \quad 0 \le j_i \le k_i \quad (1 \le i \le r). \tag{6}$$

If we multiply this by the positive integer

$$b = \prod_{i=1}^{r} p_i^{k_i - j_i},$$

we get n, so numbers of type (6) are indeed divisors of n.

The multiples of n are those numbers that have n as a divisor, so from the last results, it follows that in their canonical decomposition, multiples of n must have all the primes p_i to at least the exponent k_i $(1 \le i \le r)$. Thus the multiples are all numbers t that have canonical decomposition

$$t = \prod_{i=1}^{u} p_i^{s_i}, \quad \text{where} \quad u \ge r, \quad \text{and} \quad k_i \le s_i \quad \text{for} \quad 0 \le i \le r. \tag{7}$$

Using the decomposition from (5), we can get a formula for the number of divisors of a number. We will denote this by $\tau(n)$.[9] Each of the j_i's in (6)

---

[9] The notation $d(n)$ is also used in the literature.

can take on $k_i + 1$ values, and they can be chosen independently from each other, giving:

$$\tau(n) = \prod_{i=1}^{r}(k_i + 1). \tag{8}$$

Theorem 9 *The number n given in its canonical decomposition (5) has divisors given by (6), multiples given by (7), and the number of its divisors given by (8).*

Exercises:

  34. Which numbers have an odd number of divisors?

  35. Which numbers less than 1000 have the most divisors?

15. We need to clarify the role of 0 when considering common divisors and common multiples of numbers. We can disregard any zeros when considering common divisors, as long as not all of the numbers are zero. This is because every number is a divisor of 0. If all of the numbers are 0, then every number is a common divisor.

    Even in the case of common multiples, we can restrict ourselves to nonzero numbers. Namely, if 0 occurs among the numbers, then the only common multiple is 0, because the only multiple of 0 is 0 itself, and 0 is a multiple of every number.

    When considering common divisors and multiples, it will be advantageous to consider all occurring primes in the canonical decomposition of each of the numbers. If the prime does not divide one of the numbers, we write it with exponent 0. If $p_1, p_2, \ldots, p_r$ are the primes occurring in the decompositions of the numbers $n_1, n_2, \ldots, n_s$, we can write the canonical decompositions as

$$n_i = \prod_{h=1}^{r} p_h^{k_{ih}}, \qquad i = 1, 2, \ldots, s. \tag{9}$$

And by Theorem 9, the common divisors are the numbers

$$a = \prod_{h=1}^{r} p_h^{j_h},$$

where j_h is not larger than any of the exponents $k_{1h}, k_{2h}, \ldots, k_{sh}$. Defining m_h to be the smallest of these exponents, we can rephrase the requirement on j as follows:

$$0 \leq j_h \leq m_h, \qquad h = 1, 2, \ldots, r.$$

The numbers that we get in this way are all divisors, by Theorem 9, of

$$D = \prod_{h=1}^{r} p_h^{m_h}.$$

This leads us to a surprising result: The common divisors of our numbers are exactly the divisors of D. Of course, D is a divisor of our numbers, and hence the largest common divisor by property (1). Such a number whose divisors are exactly the common divisors of $n_1, n_2, \ldots, n_s$ we will call a *distinguished common divisor*. It is not at all obvious from the definition that a distinguished common divisor exists.[10] If both a distinguished common divisor and a greatest common divisor exist, we see that they are the same.

If all the numbers are 0, then there is no greatest common divisor. However, a distinguished common divisor still exists, namely 0, since this is a common divisor and the only number that is divisible by all the common divisors (it is the only number divisible by the common divisor 0).

Common divisors always exist, for instance 1, and if not all the numbers are 0, then there are only finitely many. In this case, there is obviously a largest, and therefore a distinguished common divisor. Soon we will prove that the greatest common divisor has the property of being the distinguished common divisor, without using the fundamental theorem. We suggest that the reader try to find such a proof before continuing. It will be clear that this will not succeed without an important new idea. However, this fact follows naturally from the fundamental theorem, as we have already seen.

16. Let us determine now the common multiples of the numbers in (9). According to Theorem 9, the p_h ($h = 1, 2, \ldots, r$) in the canonical decomposition must be present to a power at least as large as the largest among the exponents $k_{1h}, k_{2h}, \ldots, k_{sh}$. We will call this largest exponent M_h, and hence any common multiple b looks like

$$b = \prod_{h=1}^{u} p_h^{g_h}, \quad \text{where} \quad u \geq r \quad \text{and} \quad g_h \geq M_h, \quad \text{if} \quad 1 \leq h \leq r.$$

By Theorem 9, all these common multiples are multiples of the number

$$t = \prod_{h=1}^{r} p_h^{M_h}.$$

Naturally, t is also a common multiple, and therefore the smallest one. Such a number that is a multiple of the given numbers and a divisor of all of their common multiples is called a *distinguished common multiple*. The smallest common multiple is obviously well-defined and unique, hence so is the distinguished common multiple. In contrast with the distinguished common divisor, it is easy to see that the smallest common multiple is the distinguished common multiple as well (see Exercise 36). It is also interesting that from this fact the fundamental theorem can be deduced.[11]

We summarize our results in the following theorem:

---

[10] A distinguished common divisor may not exist for generalized notions of divisibility, as is seen in the settings of Exercises 27–33. See Exercise 38.

[11] Cf. K. HÄRTIG and J. SURÁNYI: *Periodica Math. Hung.* 6 (1975) pp. 235–240.

Theorem 10 *The distinguished common divisor of the numbers*

$$n_i = \prod_{h=1}^{r} p_h^{k_{ih}}, \qquad i = 1, 2, \ldots, s,$$

is

$$D = \prod_{h=1}^{r} p_h^{m_h}, \quad \text{where} \quad m_h = \min(k_{1h}, k_{2h}, \ldots, k_{sh}), \quad 1 \le h \le r;$$

the distinguished common multiple is

$$t = \prod_{h=1}^{r} p_h^{M_h}, \quad \text{where} \quad M_h = \max(k_{1h}, k_{2h}, \ldots, k_{sh}), \quad 1 \le h \le r.$$

If some of the n_i are 0, then we ignore them in the case of the distinguished common divisor, unless all of the n_i are 0, in which case the distinguished common divisor is also 0. If at least one of the n_i is 0, then the distinguished common multiple is 0.

    Both the distinguished common divisor and the distinguished common multiple are uniquely defined.

    The distinguished common divisor is equal to the greatest common divisor if the latter exists (this is the case when not all of the n_i's are 0), and the distinguished common multiple is equal to the least common multiple.

    When considering distinguished common multiples and divisors where the n_i's can be arbitrary integers, we use the absolute values of the numbers.

    From this point on, when we say greatest common divisor (or g.c.d.) and least common multiple (or l.c.m.), we will actually mean the distinguished ones, unless otherwise stated. We will use the notation

$$(n_1, n_2, \ldots, n_r) \quad \text{and} \quad [n_1, n_2, \ldots, n_r]$$

for g.c.d. and l.c.m., respectively.

    Using the notation for the g.c.d., we can easily express that $n_1, n_2, \ldots, n_r$ are relatively prime:

$$(n_1, n_2, \ldots, n_r) = 1,$$

since this means that they do not have any common divisors other than 1. We will often use this notation, but do not forget that being relatively prime is a simpler notion than g.c.d., and we may define it without using the notion of g.c.d. (for example in cases when the g.c.d. does not exist, which is possible, as we will see in the following exercises).

Exercises:

    36. Prove without using the fundamental theorem that

(a) the l.c.m. has the distinguished property,

(b*) the g.c.d. has the distinguished property.

37. Define g.c.d., l.c.m., distinguished common divisor, and distinguished common multiple in the set of polynomials with real, rational, and integer coefficients, and determine in which cases they exist. (See Exercises 22 and 23).

38. (Continuation) Solve the above exercise for the number systems introduced in Exercises 25–31.

17. We can easily decide whether numbers are relatively prime just by looking at their prime decompositions. If there is a common divisor greater than 1, then by Theorem 3, it has a prime divisor, and that is also a divisor of all the numbers; hence either it or its opposite occurs in all decompositions.

Conversely, if a prime or its opposite appears in the decomposition of every number, then this prime is obviously a common divisor. Therefore, the numbers are not relatively prime. It follows that given numbers are relatively prime if and only if there is no prime such that an associate of it appears in every decomposition.

We may also conclude that given numbers are pairwise relatively prime if and only if associates of every prime that occurs among them appear in only one decomposition.

18. In addition to the integers, a larger set of numbers plays an important role in number theory as well: the *rational numbers*. These are numbers that can be written as the quotient of two integers. (Integers are rational numbers, for example rational numbers with denominator 1.) Those numbers that are not rational are called *irrational*.

By looking at the canonical decomposition of a number, we can immediately determine whether or not it is a kth power, for any integer k greater than 1. Since if $c = n^k$, where n has a canonical decomposition

$$ep_1^{\alpha_1} p_2^{\alpha_2} \cdots p_r^{\alpha_r},$$

e is 1 or -1, and the p_i are distinct positive primes, then

$$c = e^k p_1^{k\alpha_1} p_2^{k\alpha_2} \cdots p_r^{k\alpha_r}.$$

This is exactly the canonical decomposition of c. Therefore, according to the above, every other canonical decomposition can differ at most by primes to the 0th power from this one. If k is even, then $e^k = 1$; if it is odd, then $e^k = e$. So if c is an integer other than 0, and hence has a canonical decomposition, then in this decomposition the exponent of every prime is divisible by k, and c is positive if k is even. The observations remain valid if we include primes with exponent 0 in the decomposition.

Conversely, it is clear that if the above conditions are satisfied for a given number, then it is either the kth power of an integer or the opposite of the kth power of an integer; in other words, this number is an associate of a kth power.

Based on these observations, the next theorem follows easily.

Theorem 11 *If a positive integer is not a kth power, then its kth root is irrational.*

We prove the theorem indirectly. If a positive integer n is not a kth power, then in its canonical decomposition there is a prime raised to a power not divisible by k. If $\sqrt[k]{n}$ were rational, we could write it as u/v, where u and v are integers. Then the equality

$$nv^k = u^k$$

would hold.

If we replace every factor by its canonical decomposition, then in the decompositions of u^k and v^k, every prime would appear to a power divisible by k. If on the left-hand side of the above equality we take a prime whose exponent is not divisible by k in the decomposition of n, then after combining factors of the same base its exponent will still not be divisible by k, while on the right-hand side the exponent of every prime is divisible by k. This is not possible by the fundamental theorem, and this proves the claim.

19. The fundamental theorem guarantees the existence of canonical decompositions, and this has already led us to interesting applications. The theorem, however, does not help in finding these decompositions, which becomes almost hopeless as the numbers get bigger, even with the use of (today's) supercomputers. An efficient method for finding the decomposition would be of great practical use, but at the same time would be a great danger, as we shall see in the next chapter, where we will talk about cryptography.

An interesting application is that we can give a canonical decomposition for $n!$, the product of $1, 2, \ldots, n$ (see Fact 1) for every positive integer n. We find the desired decomposition by finding the canonical decomposition of each factor separately, and then summing up the exponents of a given prime for all the factors divisible by it. Define $k(n, p)$ to be the exponent of a prime p in this decomposition. Every factor with p^t in its canonical decomposition adds t to this exponent, and the number of these is the number of factors divisible by p^t minus the number of factors divisible by p^{t+1}.

In general, up to any arbitrary positive number x, every qth number is divisible by q. The number of these is the largest integer not greater than n/q, i.e., $[n/q]$. Therefore, the number we were looking for in the previous paragraph is $[n/p^t] - [n/p^{t+1}]$, which is added t times to $k(n, p)$. We have to add these for every t for which p^t is not greater than n. In the above sum, $[n/p^t]$ occurs in the term corresponding to t with t as its coefficient, and in that of $t - 1$ with $-(t - 1)$ as its coefficient. After combining like terms, it will appear with coefficient 1. Summarizing these observations, we get the following theorem:

Theorem 12 *The canonical decomposition of $n!$ is*

$$n! = \prod_{p \leq n} p^{k(n,p)}, \quad \text{where} \quad k(n,p) = \sum_{t=1}^{r} \left[\frac{n}{p^t}\right].$$

The product is over all primes not larger than n, and $r = r(n,p)$ is such that

$$p^r \leq n < p^{r+1}. \tag{10}$$

We may write a larger upper bound here instead of this and may also include primes larger than n in the product; in the first case, their contribution to the sum is 0, and in the second, case they extend the product with factors of exponent 0.

As an example of this procedure, we determine the canonical decomposition of $12! = 479\,001\,600$:

$$k(12, 2) = [12/2] + [12/4] + [12/8] = 6 + 3 + 1 = 10;$$

$$k(12, 3) = [12/3] + [12/9] = 4 + 1 = 5; \qquad k(12, 5) = [12/5] = 2;$$

$$k(12, 7) = [12/7] = 1; \qquad k(12, 11) = [12/11] = 1.$$

Hence the decomposition we are looking for is $12! = 2^{10} \cdot 3^5 \cdot 5^2 \cdot 7 \cdot 11$. (Check the result!)

20. It is possible to express binomial coefficients using factorials (see Fact 2). These are the numbers

$$\binom{n}{j} = \frac{n!}{j!(n - j)!},$$

when n is a positive integer, and $0 \leq j \leq n$. Using canonical decompositions and without referring to any algebraic or combinatorial meanings, we can show that these numbers are always integers.

Replacing factorials by their canonical decompositions, we still have no primes greater than n. Let the exponent of such a prime p be $h = h(n, j, p)$; we have

$$h(n, j, p) = k(n, p) - k(j, p) - k(n - j, p) = \sum_{t=1}^{r} \left(\left[\frac{n}{p^t}\right] - \left[\frac{j}{p^t}\right] - \left[\frac{n - j}{p^t}\right] \right).$$

Inequality (10) gives an upper bound for the above summation.

We will show that every summand above is either 0 or 1; in this way h is a nonnegative integer, and the binomial coefficient is therefore a positive integer. With the notation $u = j/p^t$ and $v = (n-j)/p^t$ each summand has the form

$$[u+v] - [u] - [v].$$

The first term is at least $u+v-1$, and the others are not greater than u and v, respectively. Therefore, the expression is greater than -1. On the other hand, the first is not larger than $u + v$, and the others are at least $u - 1$ and $v - 1$, respectively, so the expression is smaller than 2. Therefore, h is an integer between -1 and 2, i.e., 0 or 1.

With this we not only prove our claim, but we also get the surprising result that the exponent is at most r, since it is the sum of r terms not greater than 1, where r is defined by inequality (10). This gives us the following theorem:

Theorem 13 *The canonical decomposition of the binomial coefficient $\binom{n}{j}$ is*

$$\binom{n}{j} = \prod_{p \le n} p^h, \qquad h = \sum_{t=1}^{r} \left(\left[\frac{n}{p^t} \right] - \left[\frac{j}{p^t} \right] - \left[\frac{n-j}{p^t} \right] \right),$$

where $r = r(n,p)$ satisfies $p^r \le n < p^{r+1}$.

Therefore, *the powers of primes in the above expression are not greater than n.*

Exercises:

39. Verify the following properties for the integral part function:
   (a) If n is an integer, then $[a + n] = [a] + n$.
   (b)
   $$\left[\sum_{i=1}^{k} a_i \right] \ge \sum_{i=1}^{k} [a_i].$$

   (c) If n is a positive integer, then
   $$\sum_{j=0}^{n-1} \left[a + \frac{j}{n} \right] = [na].$$

40. Show that if n is a positive integer, then
$$\left[\frac{[a]}{n} \right] = \left[\frac{a}{n} \right].$$

41. Let a be a real number not smaller than 1, and n a nonnegative integer not greater than a; show that
$$[a] > \frac{n}{n+1} a.$$

42*. With the help of the integral part function, express the following functions:

(a) the function whose value is 1 for integers and 0 for all other numbers.

(b) the smallest integer that is not smaller than the real number a (this function is called the *upper integral part of a* and is written $\lceil a \rceil$. See Fact 8).

(c) the function that gives the distance from a real number a to the closest integer.

43*. Prove that $\binom{2n}{n}$ is divisible by $n + 1$.

44*. Prove that the following number is an integer:

$$\frac{(2m)!(2n)!}{m!n!(m+n)!}.$$

21. Using the results we have obtained, we will examine a problem dating back to ancient times, namely determining the Pythagorean triples. According to the Pythagorean theorem, if the legs of a right triangle have lengths x and y, and the hypotenuse has length z, then

$$x^2 + y^2 = z^2. \tag{11}$$

Conversely, if this relation holds for three positive distances, then there is a right triangle with sides having these lengths, where z is the length of the hypotenuse.

Numbers satisfying (11) are called *Pythagorean triples*. Let us examine these more closely. Observe that x and y cannot both be odd. Namely, it is easy to see that the number preceding the square of an odd integer is divisible by 4; if n is odd, we have

$$n^2 - 1 = (n-1)(n+1),$$

and both of the factors are even, hence the product is divisible by 4.

From this observation, if both x and y were odd, their sum would be 2 greater than a number divisible by 4, but an even square is the square of an even number, and hence divisible by 4.

22. Before continuing with this investigation, we note that the above observation can be strengthened,

Theorem 14 *The number preceding the square of an odd integer is divisible by 8.*

Using the same notation, since $n-1$ and $n+1$ are both even, we can rewrite the above equation as

$$n^2 - 1 = 4\left(\frac{n-1}{2}\right)\left(\frac{n+1}{2}\right).$$

The factors in parentheses are consecutive integers; hence one of them is even and thus $n^2 - 1$ is divisible by 8.

23. Returning to (11), we have shown that at least one of x or y is even. The roles of x and y are symmetric, so we may assume that x is even. Then from (11) it follows that

$$\left(\frac{x}{2}\right)^2 = \left(\frac{z-y}{2}\right)\left(\frac{z+y}{2}\right). \tag{12}$$

Denoting the g.c.d. of $x/2$ and $(z-y)/2$ by r, for appropriate integers a and b we have

$$\frac{x}{2} = ra, \qquad \frac{(z-y)}{2} = rb.$$

Here a and b are relatively prime, since if there were a common divisor c greater than 1, then rc would be a common divisor of $x/2$ and $(z-y)/2$ greater than r, contradicting the fact that r is their g.c.d.

Rewriting (12) in terms of the new parameters and simplifying, we get

$$ra^2 = b\left(\frac{z+y}{2}\right).$$

Here a^2 and b are also relatively prime, so by Euclid's lemma b is a divisor of r, and hence there is an integer s such that $r = sb$. We may further simplify, getting

$$x = 2sab, \qquad z - y = 2sb^2, \qquad z + y = 2sa^2.$$

From this we have the following expressions for y and z:

$$y = s\left(a^2 - b^2\right), \qquad z = s\left(a^2 + b^2\right).$$

24. Since a and b are relatively prime, at least one of them is odd. If the other is even, then we will show that for $s = 1$ we get a relatively prime triple (in fact, the triple is pairwise relatively prime). If x had a common divisor greater than 1 with either y or z, it would have a common prime divisor. Such a divisor would divide either a or b, but could divide y or z only if it divided the square of the other parameter as well. Since this divisor is a prime, it would divide the base of the square as well, dividing both a and b, contradicting that a and b are relatively prime.

If y and z had a common divisor greater than 1, then again they would have a common prime divisor, and since they are both odd numbers, this cannot be 2. On the other hand, this would be a common divisor of both their sum and difference, and therefore would divide $2a^2$ and $2b^2$ as well. Since it does not divide 2, it would be a common divisor of a^2 and b^2. This would be possible only if it were a common divisor of a and b, but this is not possible.

We call a Pythagorean triple that is relatively prime a *primitive triple*.

If both a and b are odd, then there are integers a' and b' such that

$$a + b = 2a', \qquad a - b = 2b'.$$

It follows that $a = a' + b'$, and this can be odd only if one of a' or b' is even and the other odd. Using the new parameters, it follows that

$$a^2 - b^2 = 4a'b', \qquad a^2 + b^2 = 2\left(a'^2 + b'^2\right), \qquad ab = a'^2 - b'^2.$$

Replacing the old parameters by a' and b', formally the only change is that instead of s we have $2s$ and the roles of x and y are interchanged. The multipliers of $2s$ are again relatively prime, and we have reduced this case to the first one.

For every a and b, the triple given by the formulas satisfy (11), since

$$(2sab)^2 + \left(s\left(a^2 - b^2\right)\right)^2 = \left(s\left(a^2 + b^2\right)\right)^2$$

is an identity.

We summarize the above results in the following theorem:

Theorem 15 *All Pythagorean triples are of the form*

$$x = 2sab, \qquad y = s\left(a^2 - b^2\right), \qquad z = s\left(a^2 + b^2\right),$$

where s, a, and b are positive integers, a and b are relatively prime, one of them is even, and $a > b$.

The primitive triples are those for which $s = 1$.

We leave the proof of the last statement to the reader (see Exercise 45).

The condition $a > b$ is necessary for y to be positive, and can be omitted if we consider instead the absolute value of y.

Exercises:

45. Prove that in (11), if the numbers are relatively prime, then they are pairwise relatively prime and z is odd.

46. Prove that the cube of a number not divisible by 3 has a remainder of 1 or -1 upon division by 9.

47. Decide whether we obtain every Pythagorean triple if in Theorem 15 we omit the parameter s as well as the requirement that a and b be relatively prime.

48. Prove that the product of the elements of any Pythagorean triple is divisible by 60.

49. Prove that in a primitive Pythagorean triple, the number corresponding to the hypotenuse cannot be divisible by 7.

25. We can consider the results of Theorem 15 so as to describe those squares that can be written as the sum of two nonzero squares. Later we will discuss the problem of the conditions under which a positive integer can be written as the sum of (at most) two squares (see Theorem 7.2). At this time we prove only the following negative result:

Theorem 16 *If a and b are integers, then the sum*

$$a^2 + b^2$$

cannot have a positive divisor of the form $4k - 1$ relatively prime to a and b.

It follows from this theorem that primes of the form $4k - 1$ cannot be written as the sum of two squares (see Exercise 50). Later we will give a simpler proof of this theorem, but it is interesting to see that we can prove this based on the tools we already have.

We prove the theorem indirectly. The proof will be based on the following: for a given divisor of the form $4k-1$, we choose a and b as small as possible in such a way that the quotient upon division is a positive integer smaller than the divisor. This quotient will also have a divisor of the condition satisfying the theorem (of the form $4k-1$). In this way, from our assumed divisor, we get infinitely many smaller and smaller positive divisors, but this is impossible. This type of proof is an indirect analogue of induction. PIERRE DE FERMAT (1601–1665) introduced this method of proof, and favored its use. He called it *descente infinie*, or infinite descent. The proof is easier to understand if we start with the smallest possible divisor and then find a smaller one.

Assume that there exists a positive number of the form $4k - 1$ that has a multiple that is the sum of two squares, where the bases of the squares are both relatively prime to the number. Let c be the smallest such number and call the multiple $a_1^2 + b_1^2$, where

$$(c, a_1) = (c, b_1) = 1. \tag{13}$$

The numbers a_1 and b_1 are at a distance of at most $c/2$ from the closest multiple of the odd number c. Therefore, there exist integers q, r, a_2, b_2 such that

$$a_1 = cq + a_2, \quad |a_2| \leq \frac{c}{2}, \quad b_1 = cr + b_2, \quad |b_2| \leq \frac{c}{2}, \tag{14}$$

and by (13), both a_2 and b_2 are nonzero. Using this fact, we get

$$a_1^2 + b_1^2 = c\left(cq^2 + cr^2 + 2qa_2 + 2rb_2\right) + \left(a_2^2 + b_2^2\right).$$

The left-hand side and the first term of the right-hand side are divisible by c. Therefore, so is the second term on the right. Here a_2 and b_2 are both relatively prime to c, since all of the common divisors of a_2 and c also divide a_1 by (14), but a_1 is relatively prime to c, and hence so is a_2. A similar argument works for b_2 as well.

Let d be the g.c.d. of a_2 and b_2. Then we have

$$a_2 = da_3, \quad b_2 = db_3, \quad (a_3, b_3) = 1.$$

It follows from the assumption on c that $(c, d) = 1$ and $c \mid d^2 \left(a_3^2 + b_3^2 \right)$. By the last equation and Euclid's lemma, the last factor is divisible by c. It is obvious that

$$|a_3| \le |a_2|, \qquad |b_3| \le |b_2|,$$

and from (14) it follows that

$$a_3^2 + b_3^2 \le a_2^2 + b_2^2 \le \frac{c^2}{4} + \frac{c^2}{4} < c^2.$$

Summarizing our results, there is a positive integer c' for which

$$cc' = a_3^2 + b_3^2 < c^2; \qquad \text{i.e.,} \quad c' < c.$$

If we can show that c' also has a divisor of the form $4k - 1$, then we get the desired contradiction. With this in mind, let us examine the quotient upon division by 4 of the sum of squares. The square of an even number is divisible by 4, and by Theorem 14, the square of an odd number has remainder 1 even upon division by 8.

Since a_3 and b_3 are relatively prime, at least one of them is odd, and $a_3^2 + b_3^2$ is of the form $4m + 1$ or $8m + 2$, depending on whether only one is odd or both are odd. In this case the equality

$$cc' = a_3^2 + b_3^2$$

can hold only if c' (as well as c) is also of the form $4k - 1$, or if it is twice such a number.

Additionally, c' is also relatively prime to the two squares, for if it had a common divisor with one of them, this would necessarily divide the other, but a_3 and b_3, as well as their squares, are relatively prime to each other.

We conclude from this that either c' or $c'/2$ is a positive integer less than c of the form $4k - 1$ that divides the sum of two squares where the bases of the squares are relatively prime to c' or $c'/2$, respectively. This contradicts the fact that c was the smallest such number, and therefore such a number cannot exist, as is claimed by the theorem.

Exercises:

50. Prove that primes of the form $4k - 1$ cannot be written as the sum of two squares.

51. Prove that there are infinitely many primes of the form $4k + 1$.

52*. Prove that numbers of the form $3a^2 + b^2$, where a and b are integers, cannot have any positive divisors of the form $6k - 1$ that are relatively prime to a and b.

53. Prove that there are infinitely many primes of the form $6k + 1$.

26. In the sections above we used the canonical decomposition of numbers to show the existence of distinguished common divisors and common multiples as well as to determine them. These results are important, even though in practice, determining these decompositions for large numbers can be almost impossible, as we mentioned earlier.

The distinguished common divisor has, in addition, two important properties that are not easily seen from the results above. The first is that the distinguished common divisor can be expressed as the sum of integer multiples of the numbers, and the second is that if ca_i and a_i $(i = 1, 2, \ldots, r)$ are integers, then

$$(ca_1, ca_2, \ldots, ca_r) = |c| \, (a_1, a_2, \ldots, a_r).$$

We will give a new proof of the existence of the distinguished common divisor from which the above properties follow easily. We do not use the fundamental theorem, and in fact these properties provide a new proof of this theorem. Correspondingly, *we will use only the properties of divisibility from Sections 4–5 and the definitions of the required notions including that of the distinguished common divisor.* Another advantage of the new proof is that it leads to a useful method for finding the distinguished common divisor. We will prove the following theorem:

Theorem 17 *If the integers* $a_1, a_2, \ldots, a_r$ *have a distinguished common divisor, then it is unique, and can be written as*

$$\sum_{i=1}^{r} a_i u_i, \tag{15}$$

where $u_1, u_2, \ldots, u_r$ *are integers.*

If c is such that $ca_1, ca_2, \ldots, ca_r$ are integers, then the distinguished common divisor of these is $|c|$ times the distinguished common divisor of the a_i's.

If there exists a distinguished common divisor, then its uniqueness is immediate from the definition, for if D and D' are both distinguished common divisors, then because the latter is a common divisor and the former a distinguished common divisor, it follows that $D' \mid D$. Similarly, we have $D \mid D'$. By our definition, the distinguished common divisor is not negative, and therefore $D = D'$.

To prove the remaining claims of the theorem, we start from the observation that for any integers $x_1, x_2, \ldots, x_r$, the sum

$$\sum_{i=1}^{r} a_i x_i \tag{16}$$

is divisible by every common divisor of the numbers $a_1, a_2, \ldots, a_r$. Denote the set of all such numbers arising from (16) by $\mathcal{I}$. It is clear that this set contains all the a_j's, all integer multiples of any element of $\mathcal{I}$, and the sum of arbitrary elements of $\mathcal{I}$.

It is further obvious that if all the a_j's are 0, then $\mathcal{I}$ consists of only the number 0; this is the distinguished common divisor, and it satisfies the claims of the theorem.

If at least one of the given numbers is nonzero, then $\mathcal{I}$ has a positive element, because if a number is in the set, so is its opposite. Any nonempty set of positive integers has a least element. Let the least positive element of $\mathcal{I}$ be

$$m = \sum_{i=1}^{r} a_i y_i. \tag{17}$$

We claim that this will satisfy the conditions of the theorem. First, we show that every element n of $\mathcal{I}$ is divisible by m. Dividing n by m with remainder, we get

$$n = mq + s, \quad \text{where} \quad 0 \le s < m.$$

Here mq is in $\mathcal{I}$, since it is a multiple of m; hence so is $s = n - mq$. Because m is the smallest positive element of $\mathcal{I}$, the remainder s must be 0, which means that m divides n.

Therefore, m divides the a_i's, since each a_i is in $\mathcal{I}$. Since m is a member of $\mathcal{I}$, it is of the form (17), and is therefore divisible by all common divisors of the a_i's.

It remains only to prove the last statement. A distinguished common divisor of the numbers ca_i $(i = 1, 2, \ldots, r)$, by what we just proved, is the smallest positive element of the set of numbers of the form

$$\sum_{i=1}^{r} ca_i x_i.$$

This set consists of the elements that are c times an element of $\mathcal{I}$, and its smallest positive element is $|c|$ times the smallest positive element of $\mathcal{I}$. This completes the proof of the theorem.

27. It is not clear right away that the proof gives us a method for finding the distinguished common divisor, since there are infinitely many members of the set, and we cannot decide at a glance which is the smallest.

We can interpret it, however, as a method by which if we make a "blunder" in choosing, we replace our choice by a smaller number. We do not have to consider infinitely many numbers, for it is enough to start with the a_j's, since the distinguished common divisor cannot have absolute value larger than that of any of the a_j's. We can start by taking the smallest among the a_j's. We then divide the other numbers by this one; if not all of the remainders are 0, we replace our number by the smallest of the remainders, continuing this until we are done.

We explain how this method works in the simplest case of two numbers. This will also suffice for the case of more than two numbers. This already appears in EUCLID's *Elements*, and for this reason, it is called the *Euclidean algorithm.* We leave the formulation of the algorithm in the case of more than two numbers to the reader.

Let a and b be integers. If one is zero, then the other is the distinguished common divisor. If neither is zero, then we divide a by b, with remainder. If the remainder is not zero, we divide b by the remainder, then the old remainder by the new one, and so on. This procedure must terminate, since the remainders are nonnegative decreasing integers, and such a series cannot be infinite. In this way we get the following formulation:[12]

$$
\begin{aligned}
a &= bq_1 + r_1, & 0 &< r_1 < |b|, \\
b &= r_1 q_2 + r_2, & 0 &< r_2 < r_1, \\
r_{j-1} &= r_j q_j + r_{j+1}, & 0 &< r_{j+1} < r_j, & j = 2,3,\ldots,n-1, \\
r_{n-1} &= r_n q_{n+1}.
\end{aligned}
$$

It is easy to see that the last nonzero remainder r_n is the distinguished common divisor of a and b. We will prove a little more, namely that every remainder is divisible by all common divisors of a and b. Furthermore, r_n is not only a divisor of a and b, but also of all the remainders.

If $d \mid a$ and $d \mid b$, then by property (k) and the first equality, we get that $d \mid r_1$. Using this result and that $d \mid b$, from the second equality it follows that $d \mid r_2$. Proceeding in a similar fashion, if $d \mid r_{j-1}$ and $d \mid r_j$, $2 \le j \le n-1$, then the $(j+1)$st equality shows that $d \mid r_{j+1}$. The last case here is the desired conclusion, $d \mid r_n$.

To prove the inverse, we proceed from the bottom up. The last equation says that $r_n \mid r_{n-1}$. If for $n-1 \ge j \ge 2$, we have that $r_n \mid r_{j+1}$ and $r_n \mid r_j$, then again using property (k), it follows that $r_n \mid r_{j-1}$. Finally, the second and first equalities give the desired $r_n \mid b$ and $r_n \mid a$. With this we see that a and b have a distinguished common divisor, namely

$$(a,b) = r_n.$$

Exercises:

54. (a) Prove the following equality:

$$(n_1, \ldots, n_s, n_{s+1}, \ldots, n_t) = ((n_1, \ldots, n_s), n_{s+1}, \ldots, n_t).$$

(b) Describe how to use the Euclidean algorithm repeatedly to find the distinguished common divisor of more than two numbers.

55. Describe an algorithm to find the distinguished common divisor of more than two numbers all at once, as sketched above.

---

[12] The first two equations do not differ significantly from the subsequent ones, since if we define $r_{-1} = a$ and $r_0 = b$, then the first two equations correspond to $j = 0$ and $j = 1$, respectively.

56. Using the Euclidean algorithm, determine how to find a u and v that satisfy

$$(a, b) = au + bv. \tag{18}$$

We note here that there are infinitely many pairs that satisfy (18), for if it is satisfied by a pair u, v, then for every integer w, the equation is also satisfied by $u + wb$ and $v - wa$.

57. Which of the following expressions are identities? Support your answers.

(a) $[[a_1, a_2, \ldots, a_r], a_{r+1}, \ldots, a_{r+s}] = [a_1, a_2, \ldots, a_{r+s}]$,

(b) $((a, b), (a, c)) = (a, b, c)$,

(c) $[[a, b], [a, c]] = [a, b, c]$,

(d) $(a, b)(c, d) = (ac, ad, bc, bd)$,

(e) $[[a, b], [c, d]] = [ac, ad, bc, bd]$,

(f) $(a, b)[a, b] = ab$,

(g) $(ab, ac, bc)[a, b, c] = abc$,

(h) $[ab, ac, bc](a, b, c) = abc$,

(i) $((a, b), (a, c), (b, c))[a, b, c] = abc$,

(j) $[[a, b][a, c]] = a\left[\dfrac{b}{(a, b)}, \dfrac{c}{(a, c)}\right]$,

(k) $[a, b][a, c][b, c] = [[a, b], [a, c], [b, c]][a, b, c]$,

(l) $[[a, b], [a, c], [b, c]](a, b, c) = abc$,

(m) $([a, b], [a, c], [b, c])[a, b, c] = abc$,

(n) $[(a_1, \ldots, a_r, c), (b_1, \ldots, b_s, c)] = ([(a_1, \ldots, a_r), (b_1, \ldots, b_s)], c)$,

(o) $([a_1, \ldots, a_r, c], [b_1, \ldots, b_s, c]) = [([a_1, \ldots, a_r], [b_1, \ldots, b_s]), c]$.

28. Now we easily see, for example using (18), that indecomposable numbers have the prime property, and based on this we have the fundamental theorem.

Suppose that an indecomposable number c divides the product ab. Then either b is divisible by c, or b and c do not share a common divisor other than a unit, since c is indecomposable. In the latter case we have $(b, c) = 1$, and hence by Theorem 17 there exist numbers u and v for which $bu + cv = 1$.[13] Multiplying by a we get

$$abu + acv = a.$$

Both terms on the left are divisible by c (the first term by assumption), so by property (i), the sum a is also divisible by c. Thus one of the factors of the product is divisible by c; i.e., c has the prime property.

We can naturally extend the result to more factors: *If an indecomposable number is a divisor of a product, then it is a divisor of one of the factors.* We leave the proof to the reader.

---

[13] The Euclidean algorithm even provides a method for determining u and v. See Exercise 56.

Exercise:

58. Prove the claim made above.

We can prove the fundamental theorem indirectly, based on what we just did. Assume that there exists a number that can be decomposed into decomposable factors in two significantly different ways (not counting order and units):

$$p_1 p_2 \cdots p_r = q_1 q_2 \cdots q_s.$$

We can simplify the above equation if we divide by associates that appear on each side (we can incorporate any -1 we may get into the first factor). Suppose that this has already been done. The fact that the two decompositions are significantly different means that after this simplification we still have indecomposable factors on both sides.

This is not possible, because p_1, for example, would be a divisor of the product on the right, which means that it would have to divide one of its factors, q_j. But the latter is indecomposable, meaning that it does not have any other divisors besides its associates and the units. This means that p_1 and q_j are associates, but in this case we would have already canceled them out.

We have arrived at a contradiction. Therefore, there is no number that has two significantly different decompositions into indecomposables.

In the above we proved only what was needed in the proof of the fundamental theorem, but in the same way we can prove Euclid's lemma (Theorem 6) as well.

Exercises:

59. Prove that indecomposable numbers have the prime property, using the following relation:

$$(ca_1, ca_2, \ldots, ca_s) = |c| \, (a_1, a_2, \ldots, a_s),$$

for ca_i integers, $i = 1, 2, \ldots, s$.

60. Prove Theorem 6 using the properties of the g.c.d.

29. We now turn to a natural application. We assume everything we have done so far, using the statements of Theorem 17 often. Problems in which we look for integer solutions are called *Diophantine*.[14] For instance, Pythagorean triples are solutions to the Diophantine equation (11). Diophantine equations are usually hard, and rarely do we get such nice solutions as in this case. We can expect that first-order Diophantine problems do not belong to this group.

---

[14] Named after DIOPHANTUS OF ALEXANDRIA, the Greek mathematician from the third century who studied first- and second-degree equations of this type.

We will deal only with equations of this type that have two unknowns. Given integers a, b, and c, find integers x and y satisfying

$$ax + by = c. \qquad (19)$$

Is it possible that this does not have a solution? Naturally, the answer is yes. Consider the equation $12x + 18y = 25$. For all integers x and y, the left-hand side is always even, and hence it is impossible to get 25. (We cannot even get 26, since the left-hand side is divisible by 6.) In general, for all integers x and y, the left-hand side is divisible by all common divisors of a and b, including (a, b). Therefore (19) can have integer solutions only if

$$(a, b) \mid c. \qquad (20)$$

If this holds, meaning that there is an integer d for which

$$c = (a, b)\, d,$$

then multiplying (18) by d we have

$$c = (a, b)\, d = aud + bvd,$$

yielding $x = ud$ and $y = vd$ as solutions of (19). The problem, therefore, has solutions if and only if (20) holds.

30. The Euclidean algorithm is actually very useful in practice for finding (a, b) and an appropriate u and v. We know that there are infinitely many such pairs of numbers u and v. We now determine all solutions to (19). Assume that x_0, y_0 and x', y' satisfy the equations

$$ax_0 + by_0 = c \quad \text{and} \quad ax' + by' = c.$$

Subtracting the second equation from the first, we get

$$a(x_0 + x') + b(y_0 - y') = 0.$$

We divide this by (a, b), rearranging terms to get

$$\frac{a}{(a, b)}(x_0 - x') = \frac{b}{(a, b)}(y' - y_0),$$

The first factors of the two sides are relatively prime, because by the last claim in Theorem 17,

$$\left(\frac{a}{(a, b)}, \frac{b}{(a, b)} \right) = \frac{1}{(a, b)}(a, b) = 1.$$

On the other hand, $a/(a, b)$ is a divisor of the product on the right-hand side, and hence by Euclid's lemma, it is also a divisor of the second factor, so there exists an integer w for which

$$y' - y_o = \frac{wa}{(a,b)}, \quad \text{or} \quad y' = y_0 + \frac{wa}{(a,b)}.$$

Substituting this and solving for x', we have

$$x' = x_0 - \frac{wb}{(a,b)}.$$

It is easy to see by substitution that for every number w (including nonintegers), the resulting pair x', y' is a solution to (19). We summarize our results in the following theorem:

Theorem 18 *The equation*

$$ax + by = c$$

has a solution if and only if $(a,b) \mid c$. If this is satisfied and x_0, y_0 is a solution to the equation, then all solutions are given by the formula

$$x' = x_o - \frac{wb}{(a,b)}, \quad y' = y_o + \frac{wa}{(a,b)},$$

where w is any integer.
    We can find a solution with the help of the Euclidean algorithm.

Exercises:

61. Give necessary and sufficient conditions for the existence of integer solutions to the following equation:

$$a_1x_1 + a_2x_2 + \cdots + a_nx_n = b,$$

where the a_i's and b are given integers.

62. (Continuation) Give an algorithm (if possible more than one) to find a solution to the above equation when a solution exists.

63. In how many ways can we pay one dollar using nickels, dimes, and quarters?

64. (a) Prove that for every integer a and b, the following system of equations has an integer solution:

$$x + y + 2z + 2t = a,$$

$$2x - 2y + z - t = b.$$

(b) Determine all solutions to the above system of equations.

65*. (a) For which positive integers a and b is the following true: There exists a number N such that for any integer n greater than N, there are positive integer solutions to the equation

$$ax + by = n.$$

(b) What is the value of the smallest such N?

(c) Determine the number of positive integers n for which the equation has no positive integer solutions.

With respect to this last problem, we note that for more than two unknowns there is no known expression for the smallest limit. Such a result would be very useful in terms of applications, and serious research is taking place in this area.

2. Congruences

1. The exercises in Section 1.2 were concerned with divisibility properties. In many cases rules were studied for a given divisor that gave a new number significantly smaller than the dividend, both having the same remainder upon division by the divisor (e.g., for divisibility by 2, 3, 4, 5, 8, 9, 10, and 11). In general, this relation that two numbers have the same remainder upon division by a given divisor proves to be such a useful tool in number theory that GAUSS introduced a special notation to represent it that quickly became part of almost every mathematician's working vocabulary. Taking the names from Latin, we call two such numbers *congruent*, and their divisor the *modulus*.

We say that a *is congruent to* b *modulo* m (or just mod m) if a and b have the same remainder when divided by m (for example in the sense of Theorem 1.1), or, by an equivalent statement, if $a - b$ *is divisible by* m. We write this $a \equiv b \pmod{m}$. Two numbers that are not congruent are called *incongruent*. We write this as $a \not\equiv b \pmod{m}$.

We can now express divisibility with the help of congruence:

$$a \equiv 0 \pmod{m}$$

that means the same as $m \mid a$.

We usually require the modulus to be an integer greater than 1, but the definition works for all integers. Negative numbers can be ignored, since congruence to modulus m and congruence to modulus $-m$ are the same by property (h).

Congruence modulo 0 is the same as equality, since $a \equiv b \pmod{0}$ means that $a = b$, by property (e), and it is unnecessary to introduce a more complicated notation for this.

Congruence modulo 1 is also not very useful, since in that case all integers are congruent.[1]

2. The fact that congruence has properties similar to equality makes it very useful. It is clear (so clear that we leave the proof of the following properties to the reader) that congruence is a reflexive, symmetric, and transitive relation; i.e., for all integers a, b, c, and m, the following properties hold:

---

[1] For the real numbers, this does have meaning. It says that the fractional parts of the two numbers are equal, and this notation is often used.

(a) $a \equiv a \pmod{m}$,

(b) if $a \equiv b \pmod{m}$, then $b \equiv a \pmod{m}$,

(c) if $a \equiv b \pmod{m}$ and $b \equiv c \pmod{m}$, then $a \equiv c \pmod{m}$.

Regarding basic operations, if $a \equiv b \pmod{m}$, then

(d) $a + c \equiv b + c \pmod{m}$,

(e) $a - c \equiv b - c \pmod{m}$,

(f) $ac \equiv bc \pmod{m}$.

In the case of multiplication, we still have congruence if we additionally multiply the modulus:

(g) If $a \equiv b \pmod{m}$ then $ac \equiv bc \pmod{mc}$.

For simplification, there is already an important difference. Property (g) can be reversed:[2]

(h) If $ac \equiv bc \pmod{mc}$ and $c \neq 0$, then $a \equiv b \pmod{m}$.

A blind reversal of property (f) can easily lead to an untrue statement. For example,

$$51 \cdot 11 = 561 \equiv 1071 = 51 \cdot 21 \pmod{15},$$

but 11 and 21 are not congruent modulo 15. In general,

$$ac \equiv bc \pmod{m}$$

means that $m \mid (a-b)c$, and it does not follow that $m \mid (a-b)$, since m could split into factors dividing $a - b$ and c, respectively, as in the above example. We can exclude this case using Euclid's lemma (Theorem 1.6), and then we can simplify:

(i) If $ac \equiv bc \pmod{m}$ and $(c, m) = 1$, then $a \equiv b \pmod{m}$.

Furthermore, adding, subtracting, and multiplying congruent numbers (with respect to the same modulus) maintains congruence. In other words, if $a \equiv b \pmod{m}$ and $c \equiv d \pmod{m}$, then

(j) $a + c \equiv b + d \pmod{m}$,

(k) $a - c \equiv b - d \pmod{m}$,

(l) $ac \equiv bd \pmod{m}$.

These properties are not difficult to see on their own, but they also follow easily from properties (d), (e), and (f), respectively. For example, it follows by property (d) that $a + c \equiv b + c \pmod{m}$ and $c + b \equiv d + b \pmod{m}$, and then by transitivity we have property (j).

---

[2] The case $c = 0$ is trivial; we could have excluded it in properties (d)–(g).

We can extend these properties in the case of more than two congruences as well; by repeatedly using property (l) with the same congruence, we get the following:

(m) If $a \equiv b \pmod{m}$ and n is a positive integer, then $a^n \equiv b^n \pmod{m}$.

This last property also follows from the following well-known identity, valid for all integers $n > 1$:

$$a^n - b^n = (a - b)\left(a^{n-1} + a^{n-2}b + \cdots + b^{n-1}\right).$$

Repeatedly using the above properties, we also have the following:

(n) If $f(x)$ is a polynomial with integer coefficients, and $a \equiv b \pmod{m}$, then

$$f(a) \equiv f(b) \pmod{m}.$$

Sometimes it is useful to reduce a congruence to a simpler one, for example decreasing the modulus by a divisor, as exhibited by the following:

(o) If $a \equiv b \pmod{mm'}$ and $m' \neq 0$, then $a \equiv b \pmod{m}$.

3. We can prove the divisibility rules for 9 and 11 using congruences. We see that

$$10 \equiv 1 \pmod 9, \qquad 10 \equiv -1 \pmod{11},$$

so by property (m),

$$10^n \equiv 1 \pmod 9, \qquad 10^n \equiv (-1)^n \pmod{11}.$$

Applying these congruences and property (f) repeatedly, we get for integers $a_1, \ldots, a_n$,

$$\sum_{i=0}^{n} a_i 10^i \equiv \sum_{i=0}^{n} a_i \pmod 9, \qquad \sum_{i=0}^{n} a_i 10^i \equiv \sum_{i=0}^{n} (-1)^i a_i \pmod{11}.$$

These formulas furnish exactly the divisibility rules from the previous chapter.

In principle, we could deduce divisibility rules for every divisor in this way, but in practice, this would generally be too complicated. Let us see what this leads to in the case of 7:

$$10 \equiv 3 \pmod 7, \qquad 10^2 \equiv 2 \pmod 7, \qquad 10^3 \equiv -1 \pmod 7,$$

$$10^4 \equiv -3 \pmod 7, \qquad 10^5 \equiv -2 \pmod 7, \qquad 10^6 \equiv 1 \pmod 7.$$

From the last congruence and property (m), we get

$$10^{6k} \equiv 1 \pmod 7,$$

and using the other congruences and property (l) we have

$$10^{6k+1} \equiv 3 \pmod 7, \quad 10^{6k+2} \equiv 2 \pmod 7, \quad 10^{6k+3} \equiv -1 \pmod 7,$$
$$10^{6k+4} \equiv -3 \pmod 7, \quad 10^{6k+5} \equiv -2 \pmod 7.$$

Using this, we get the following relation:

$$\sum_{i=0}^{n} 10^i a_i \equiv \sum_{k=0}^{n} a_{6k} + 3 \sum_{k=0}^{n} a_{6k+1} + 2 \sum_{k=0}^{n} a_{6k+2}$$
$$- \sum_{k=0}^{n} a_{6k+3} - 3 \sum_{k=0}^{n} a_{6k+4} - 2 \sum_{k=0}^{n} a_{6k+5} \pmod 7.$$

Here we take all a_i for $i > n$ to be 0.

We do not even try to formulate the divisibility rule we just found, since it would be too difficult to remember, and the rule from Exercise 1.6 (a), Chapter 1, is much more useful. (It is true that the new numbers arising from that rule are not congruent to the original number modulo 7, unless the number is divisible by 7, but this is enough to decide whether the number is divisible by 7 or not.)

4. Let us put the integers into classes according to an integer m greater than 1, placing integers that are congruent to each other modulo m into one class. Based on properties (a)–(c), it is easy to see that if we choose an arbitrary element from a class, then the class consists of all elements that are congruent to it, and no two distinct classes have a common element. We call these classes the *residue classes modulo m*, and we denote them by $(a)_m$, where a is an element of this class. A residue class is therefore an arithmetic progression with difference m, infinite in both directions.

According to properties (j)–(m), addition, subtraction, multiplication, and exponentiation to the nth power are valid operations between these classes, in the sense that if we perform the operation on any representatives of two classes, the resulting element is always in the same class.

This is, however, not valid for all operations, for instance greatest common divisor and least common multiple. These are operations between the numbers themselves, but are not operations between the classes; for instance,

$$4 \equiv 39 \pmod{35}, \quad 6 \equiv 111 \pmod{35},$$

but

$$(4, 6) = 2 \quad \text{and} \quad (39, 111) = 3$$

are not congruent modulo 35, and neither are

$$[4, 6] = 12, \quad \text{and} \quad [39, 111] = 1443.$$

We note, however, that we can define a unary operation based on the greatest common divisor, which is the greatest common divisor of the residue class with the modulus. Indeed, if $a \equiv b \pmod{m}$ and $(b, m) = d$, then by property (o), $a \equiv b \equiv 0 \pmod{d}$, i.e., $d \mid a$. Since $(b, m) \mid m$, we have that

$$(b, m) \mid (a, m),$$

and from property (b), it also follows that

$$(a, m) \mid (b, m).$$

From the two relations, we conclude that

$$(a, m) = (b, m),$$

as claimed.

In particular, those classes that are relatively prime to the modulus (i.e., every element of the class is relatively prime to the modulus) will play an important role.

5. In the previous section we used property (o). We can rephrase this as follows: If two integers are in the same residue class modulo mm', then they are in the same residue class modulo m. The converse, however, is not true; if $m' > 1$, the elements from the modulo m residue classes must be spread out among the modulo mm' residue classes if only because of the greater number of classes. Let us examine how $(a)_m$ is distributed over the modulo mm' residue classes. The residue class $(a)_m$ consists of numbers of the form $a + mu$, where u runs over all integers. Let us divide u with remainder by m':

$$u = m'v + r, \quad \text{where} \quad v \quad \text{is an integer and} \quad 0 \leq r \leq m' - 1.$$

The residue class $(a)_m$, therefore, is the set of numbers of the form $a + mm'v + mr$, where r is one of $0, 1, \ldots, m' - 1$, and v is any integer. In this way we have the following congruence property:

(p) A residue class modulo m splits into m' residue classes modulo mm': $(a)_m$ splits into the residue classes $(a + mr)_{mm'}$, where $r = 0, 1, \ldots, m' - 1$.

6. We have seen that a residue class is uniquely defined by one of its elements. We can therefore choose a representative from each class and perform operations on the classes using these elements; for the result, we take the element representing the class of the result.

We call a set that contains exactly one element from every residue class modulo m a *complete residue system modulo m*.

The two remainder theorems (Theorems 1.1 and 1.1') give two special complete residue systems; the first gives the *smallest nonnegative remainders* and the second the *remainders with smallest absolute value*. We will often use these two residue systems, but we are not obliged to choose the representatives in such a nice way; for instance, $77, 100, -21, 2, -59, 666$ is a complete residue system modulo 6.

A complete residue system modulo m has two important properties: It has m elements, and the elements are pairwise incongruent modulo m. These two properties actually characterize a complete residue system.

Theorem 1 *If a set of numbers has m elements that are pairwise incongruent modulo m, then the set is a complete residue system modulo m.*

In fact, because the elements are pairwise incongruent, no two can be in the same residue class, and since there are m of them, every residue class must contain one of them.

With the help of this modest-looking theorem, we can get a new residue system from a given one, which leads to an important relation. In particular, we have the following theorem.

Theorem 2 *If a and m are relatively prime integers, b is an arbitrary integer, and $r_1, r_2, \ldots, r_m$ is a complete residue system modulo m, then*

$$ar_1 + b, ar_2 + b, \ldots, ar_m + b$$

is a complete residue system modulo m.

The system has m elements, so by Theorem 1 it is enough to show that the new numbers are pairwise incongruent. If

$$ar_i + b \equiv ar_j + b \pmod{m},$$

then by property (e) we can subtract b from both sides. Using the fact that $(a, m) = 1$ and property (i), we can simplify, eliminating a, to get

$$r_i \equiv r_j \pmod{m}.$$

Because r_i and r_j are elements of a complete residue system, this is possible only if $i = j$. Therefore, two elements can be congruent only if they are identical; hence the elements are pairwise incongruent.

Exercises:

1. Prove properties (a)–(o).

2. What are the possible last digits of fourth powers in the base-10 number system?

3. What is the last digit of 7^{888} in the base-10 system? What are the last two digits? What are the last two digits of 7^{890}?

4. Prove that a residue class is made up of those numbers that are congruent to an arbitrary representative and that different classes do not have a common element.

5. (a) Can the set of numbers $\{-17, -13, 9, 8\}$ be extended to a complete residue system modulo 21? How about $\{-20, -10, 2, 20\}$?

(b) Let m be an integer greater than 1. Give necessary and sufficient conditions for a set $\{r_1, \ldots, r_k\}$ to be extendable to a complete residue system modulo m. (Prove your answer.)

6. Let a and b be relatively prime integers. Further, let $r_1, \ldots, r_a$ be a complete residue system modulo a, and $s_1, \ldots, s_b$ a complete residue system modulo b. Do the numbers $as_i + br_j$ $(i = 1, \ldots, b; \; j = 1, \ldots, a)$ form a complete residue system modulo ab? How about the numbers $as_i + r_j$ $(i = 1, \ldots, b; \; j = 1, \ldots, a)$?

7. We mention some problems related to residue classes before turning to applications of Theorem 2.

The formation of residue classes modulo m provides a distribution of all integers into classes with no common element. If we consider residue classes all corresponding to different moduli, we do not find classes satisfying both properties above. We can, however, find disjoint classes, and also classes whose union is all integers. In the first case, we call the system of congruences *disjoint*, and in the second case *covering*.[3]

Let us examine first the case of disjoint systems of congruences. Denote by $k(x)$ the maximum number of moduli of a disjoint system, where the largest modulus is not greater than x.

For an integer d greater than 1, the classes $(0)_d, (1)_{2d}, \ldots, (d-1)_{d^2}$ do not have a common element, since dividing by d already gives a different remainder for each class. For a given x, choosing d to be the greatest integer less than $\sqrt{x}$, we get that $k(x) \geq d > \sqrt{x} - 1$.

P. ERDŐS and S. K. STEIN gave a function Φ for which

$$k(x) > \frac{x}{\Phi(x)}$$

and Φ grows more slowly than x to any positive power. They conjectured that $k(x)/x$ tends to 0 as x goes to infinity, a conjecture that was later proved by ERDŐS and SZEMERÉDI.[4] Further results and problems are included in their article.

8. We know even less about covering systems of congruences. The set of congruence classes $(0)_2, (0)_3, (1)_4, (5)_6, (7)_{12}$ is an example of a covering system. We can verify this using property (p), since the above classes split into the following residue classes modulo 12:

$$(0)_{12}, (2)_{12}, (4)_{12}, (6)_{12}, (8)_{12}, (10)_{12},$$

$$(0)_{12}, (3)_{12}, (6)_{12}, (9)_{12},$$

[3] For the origins of these types of questions, see P. ERDŐS: *Summa Brasiliensis Math. II* (1950), pp. 113–123.

[4] P. ERDŐS, E. SZEMERÉDI: *Acta Arithmetica* **15** (1968), pp. 85–90.

$$(1)_{12}, (5)_{12}, (9)_{12},$$
$$(5)_{12}, (11)_{12},$$
$$(7)_{12},$$

and among these, every residue class is represented.

There are more such systems for which the smallest modulus is 2. The following system, which has smallest modulus 3, was proposed by H. DAVENPORT and P. ERDŐS:

$$(0)_3, (0)_4, (0)_5, (1)_6, (6)_8, (3)_{10}, (5)_{12}, (11)_{15}, (7)_{20},$$
$$(10)_{24}, (2)_{30}, (34)_{40}, (59)_{60}, (98)_{120}.$$

Covering systems are known for which the smallest modulus is 4, 6 (D. SWIFT), 8 (J. L. SELFRIDGE), 9 (R. F. C. CHURCHHOUSE), and 20 (S. L. G. CHOI). We do not know whether for an arbitrary m_0 there exists such a covering system with smallest modulus m_0. To date, we do not even know the answer for all m_0 less than 20, nor do we know whether there exists a covering system for which every modulus is odd.[5]

It is easy to see for a covering system that the sum of the reciprocals of the moduli is at least 1 (see Exercises 9 and 10). The fact that the sum is strictly greater than 1 was proved by MIRSKY and NEWMANN (and independently by DAVENPORT and RADO) with the help of calculus. We present this proof below. Later, a proof was found that does not use limits; we present this proof in Section 10. We cannot expect to improve upon this result, as shown by the example

$$\left(2^{t-1}\right)_{2^t}, \quad 1 \le t \le u, \quad (2^u)_{3 \cdot 2^{u-2}}, \quad \left(2^{u+1}\right)_{3 \cdot 2^{u-1}}, \quad (0)_{3 \cdot 2^u}.$$

This is a covering system; the sum of the reciprocals of the moduli is $1 + \left(\frac{1}{3} \cdot 2^{u-2}\right)$, which, for u large enough, can be made arbitrarily close to 1. DAVENPORT conjectured that if the smallest modulus is larger than 2, then it is possible to give a lower bound greater than one, but this is still an open problem.

Exercises:

7. Find a covering system that includes the residue classes $(1)_2$ and $(2)_3$.

8. Find a covering system that does not include 6 among the moduli.

9. Prove that the sum of the reciprocals of the moduli in a covering system is at least 1.

10. Prove that if two residue classes in a covering system have a common element, then they have a common arithmetic progression, and hence the sum of the reciprocals of the moduli is greater than 1.

---

[5] See the article by P. ERDŐS referred to in footnote 3, and also S. L. G. CHOI: *Mathematics of Computation* **25** (1971) pp. 885–895.

9*. At the beginning of Section 2.7 we stated that we cannot find disjoint residue classes with distinct moduli whose union is all the integers. This will be a consequence of the following theorem.

Theorem 3 *In a covering system of congruences, the sum of the reciprocals of the moduli is larger than* 1.

The proof of this theorem uses Exercises 9 and 10, whose proofs were left to the reader. We will also use complex numbers, the formula for sums of infinite geometric progressions (see Fact 4), and EULER's ingenious idea to draw conclusions for sequences from the properties of their generating functions. For a sequence (finite or infinite) of integers $c_1, c_2, \ldots$, the corresponding function $z^{c_1} + z^{c_2} + \cdots$ is called the *generating function of the series*[6] (cf. Section 1.2, Exercise 9).

By Exercise 9 the sum of the reciprocals is at least 1. According to Exercise 10, equality is possible only if every number belongs to only one residue class. It remains only to prove that in the case of distinct moduli this sum must still be larger than 1. We prove this indirectly.

Assume that $(a_j)_{m_j}, j = 1, 2, \ldots, k$, is a covering congruence system, $m_1 < m_2 < \cdots < m_k$, and every integer belongs to only one residue class. We choose the a_j's such that $0 \leq a_j \leq m_j - 1$ $(j = 1, 2, \ldots, k)$. Then the union of the arithmetic progressions $a_j, a_j + m_j, a_j + 2m_j, \ldots, a_j + \ell m_j, \ldots$ gives every nonnegative integer exactly once, so their union can be considered as the arithmetic progression having $a_0 = 0$ and $m_0 = 1$.

For $j = 0, 1, \ldots, k$, we define the generating functions

$$F_j(z) = z^{a_j} + z^{a_j + m_j} + z^{a_j + 2m_j} + \cdots .$$

These are convergent geometric series for every $|z| < 1$.

By assumption, it is true that

$$F_1(z) + F_2(z) + \cdots + F_k(z) = F_0(z).$$

Using the formula for the sum of an infinite geometric series, we have

$$\sum_{j=1}^{k} \frac{z^{a_j}}{1 - z^{m_j}} = \frac{1}{1 - z}. \tag{1}$$

After eliminating the denominators we have

$$(1 - z) \sum_{j=1}^{k} z^{a_j} \prod_{n=1}^{k}{}^{*} (1 - z^{m_n}) = \prod_{j=1}^{k} (1 - z^{m_j}),$$

---

[6] In addition to the series $1, x, x^2, \ldots$, we can take another function sequence $f_1, f_2, \ldots$ and create generating functions of the form $c_1 f_1 + c_2 f_2 + \cdots$, as EULER himself did. Generating functions have proven to be invaluable tools in different branches of mathematics.

where the $*$ in the product sign means that we omit $n = j$.

The relation is true for every z with absolute value less than 1; hence for infinitely many values the two polynomials are equal. From this it follows that equality holds for all z.

We will show that equality does not hold for

$$z_0 = e^{2\pi i/m_k} = \cos(2\pi/m_k) + i\sin(2\pi/m_k)$$

(where i is the imaginary unit). The last factor on the right-hand side is 0 for $z = z_0$; hence the whole product is 0. With the exception of the last summand on the left-hand side, every summand is 0 too, since the last factor is 0. The value of the last summand, however, is

$$\left(1 - e^{2\pi i/m_k}\right) e^{2\pi a_k i/m_k} \prod_{n=1}^{k-1} \left(1 - e^{2\pi m_n i/m_k}\right).$$

All of the factors are nonzero, hence so is their product. This contradicts the equality, and therefore finishes the proof of the theorem.

The solution to Exercise 10 yields the further result that if two residue classes, say those of numbers u and v, have a common element, then the sum of the reciprocals of the moduli is at least $1 + 1/[m_u, m_v]$. Thus we have shown that the sum of the reciprocals of the moduli for the above covering system is at least

$$1 + \frac{1}{m_{k-1} m_k}.$$

The conjecture mentioned in Section 8 proposes that for $m_1 > 2$, the sum can be bounded away from 1 by a constant depending only on m_1.

10. A. LIFSIC gave an elementary solution to a contest problem[7] that turned out to be equivalent to Theorem 3.

Based again on Exercises 9 and 10, it is sufficient to prove that it is not possible to cover the integers by finitely many distinct residue classes, or equivalently by finitely many arithmetic progressions having distinct differences in such a way that no two of them share a common element.

We prove the statement indirectly. Assume that there is such a covering and let d be the least common multiple of the differences. Wind the number line around a circle of circumference d. On this circle, the integers represent the vertices of a regular d-sided polygon. The sum of all d vectors from the origin to the d vertices is $\mathbf{0}$. The arithmetic progressions form the vertices of disjoint regular polygons that together cover all vertices of the d-sided polygon. Let d_1 be the smallest number of sides in one of these polygons.

---

[7] LIFSIC's proof appears as the solution to Problem 143, p. 43 and pp. 159–161 in N. B. VASIL'EV, A. A. EGOROV: *Zadachi vsesoyuznikh matematicheskikh Olimpiad* (*Problems of the All-Soviet-Union Mathematical Olympiads*), NAUKA, MOSCOW, 1988 (IN RUSSIAN).

Rotate each vector around the origin by multiplying its angle by d_1. The resulting points cover a d/d_1-sided regular polygon a number of times, and the sum of the resulting vectors is still $\mathbf{0}$, since $d/d_1 \neq 1$ (there is more than one arithmetic progression). The covering polygons that do not have d_1 sides go through a polygon with more than one vertex. Thus the sum of the vectors from the origin to their vertices is $\mathbf{0}$. On the other hand, all the vertices of the d_1-sided polygon rotate into one point on the circle; hence the sum of these vectors must be a vector different from $\mathbf{0}$. This is a contradiction, and hence the assumption of the existence of such a covering must be false, and with this we have proved Theorem 3.

11. In connection with congruences, we also have congruences with unknowns that we would like to solve, similarly to algebraic equations. The easiest among these are of course those of degree 1:

$$ax \equiv b \pmod{m}. \tag{2}$$

If x_0 is a solution of the congruence, then every element of the residue class $(x_0)_m$ is a solution, so we are interested in how many residue classes satisfy the relation. It is clearly enough to check all the elements of a complete residue system.

Congruences of this type do not always have solutions. For example, in the congruence

$$21x \equiv 8 \pmod{39},$$

the left-hand side is always in a residue class where the numbers are divisible by 3. This is because 39 is also divisible by 3, and in a residue class, every element has the same greatest common divisor with the modulus. Thus the residue class $(21x)_{39}$ does not contain 8 for any x.

Exercises:

11. Which of the congruences below have solutions? For those that have solutions, find a solution.
    (a) $51x \equiv 29 \pmod{52}$,
    (b) $42x \equiv 58 \pmod{87}$,
    (c) $77x \equiv 91 \pmod{238}$,
    (d) $42x \equiv 57 \pmod{171}$.

12. Solve the following congruences:
    (a) $3^8 x \equiv 89 \pmod{100}$,
    (b) $11^3 x \equiv 47 \pmod{100}$.

13. Prove that there is a number x that is a solution to the congruence

$$\left(9k^2 + 3k + 1\right) x \equiv 111 \pmod{6k + 1}$$

for all positive integers k.

12. In the special case where $(a, m) = 1$ it easily follows from Theorem 2 that (2) always has a solution and that the solution is one residue class. This can be seen in the following way: We start with a complete residue system $r_1, r_2, \ldots, r_m$ and form a new system $ar_1, ar_2, \ldots, ar_m$. By Theorem 2, this is also a complete residue system, and therefore has exactly one element that is congruent to b modulo m.

We will reduce the general case to this special one.

If (2) is satisfied for some x and $(a, m) = d$, then by property (o) the congruence $ax \equiv b \pmod{d}$ is satisfied. But $d \mid a$ and hence the left-hand side is congruent to 0 modulo d. The congruence, therefore, can have a solution only if $d \mid b$.

Now if this condition is satisfied, then the congruence always has a solution. Dividing a, b, and m by d, we get

$$a = a'd, \quad b = b'd, \quad m = m'd, \quad \text{and} \quad (a', m') = 1, \quad \text{since} \quad d = (a, m).$$

Rewriting (2), we have

$$a'dx \equiv b'd \pmod{m'd}.$$

This congruence can be reduced to the following, using property (h):

$$a'x \equiv b' \pmod{m'}.$$

This last congruence is the special case we discussed above, and its solution is one residue class modulo m'. From property (p) there are $d = (a, m)$ solutions of (2) (counting the number of residue classes modulo m). We summarize our observations in the following theorem.

Theorem 4 *For integers a, b, m, the congruence*

$$ax \equiv b \pmod{m}$$

has solutions if and only if
$$(a, m) \mid b.$$

If this holds, then there are (a, m) residue classes modulo m that satisfy the congruence.

This theorem naturally includes the special case discussed at the beginning of this section.

13. In the previous section we gave conditions for the existence of solutions. The problem still remains of how to find these solutions. The above section yields only the trivial hint that if we try all elements of a complete residue system, we will eventually find the solutions. In the examples we saw that with a little cunning we were able to find the solutions more easily, but unfortunately, this does not hint at a better method for the general case.

There are, however, more efficient methods for finding solutions. We can find such a method by rephrasing the problem as a linear Diophantine equation of two unknowns, and in Chapter 1 we found a method for solving these using the Euclidean algorithm.

If the congruence in (2) is solvable, then there exists a y such that

$$ax - b = my, \qquad \text{i.e.,} \qquad ax - my = b,$$

and conversely, any solution to the latter equation yields an x that is a solution of the congruence. In this setting we can rephrase the statement of Theorem 1.18 to congruences of degree 1. This gives us the following theorem:

Theorem 5 *The congruence* $ax \equiv b \pmod{m}$ *has a solution if and only if the Diophantine equation*

$$ax - my = b$$

has a solution, and this has a solution if and only if

$$(a, m) \mid b.$$

If this holds, then the solution of the congruence is a residue class modulo $m/(a,m)$; *the solution can be found using the Euclidean algorithm with* a *and* m.

In Section 25 we will see a completely different method for solving the special case $(a, m) = 1$ of the congruence (2).

14. Certain problems arise in connection with linear congruences that are somewhat different from those concerning equations. We often find such questions in problems sections of magazines where the object is to determine a number having specified remainders upon division by certain divisors, which means congruences of the form $x \equiv b_i \pmod{m_i}$. We will see that if a solution to this type of problem exists, then it is a residue class modulo an appropriate modulus. With further restrictions, for example on the size of the solution, it can be made unique.

In what follows we will consider the more general system of congruences

$$a_i x \equiv b_i \pmod{m_i}, \quad i = 1, 2, \ldots, k. \tag{3}$$

We call these *simultaneous systems of congruences*. We will consider in detail only the special case where moduli are pairwise relatively prime.

It is natural to assume that each of the congruences is solvable, and additionally that

$$(a_i, m_i) = 1, \quad i = 1, 2, \ldots, k, \tag{4}$$

since we have seen that first-degree congruences can be reduced to this special case.

However, fulfillment of the conditions (4) does not guarantee the existence of a solution of the system (3), as seen in the following example:

$$x \equiv 3 \pmod{28}, \qquad x \equiv 12 \pmod{74}.$$

The first congruence can be satisfied only by odd numbers, and the second only by even numbers; hence the system has no solution.

If, however, such a system has a solution, then it is easy to see what all the solutions look like.

Theorem 6 *If the system* (3) *has a solution and the conditions* (4) *are satisfied, then all the solutions form a residue class modulo the least common multiple of the moduli from* (3).

We will first show that if x_1 and x_2 are solutions, then they must be in the same residue class. Second, we will show that if one element of a residue class is a solution, then so are all elements of this class.

Assume that x_1 and x_2 are solutions of the system (3), meaning that they satisfy the following congruences:

$$a_i x_1 \equiv b_i \pmod{m_i} \quad \text{and} \quad a_i x_2 \equiv b_i \pmod{m_i}, \quad i = 1, 2, \ldots, k.$$

For each i we consider the difference of the two congruences:

$$a_i(x_1 - x_2) \equiv 0 \pmod{m_i}, \qquad i = 1, 2, \ldots, k.$$

We can divide the ith congruence by a_i, using the conditions (4) and property (h), yielding

$$x_1 - x_2 \equiv 0 \pmod{m_i}, \quad \text{i.e.,} \quad m_i \mid x_1 - x_2, \quad i = 1, 2, \ldots, k.$$

Since $x_1 - x_2$ is divisible by the moduli $m_1, m_2, \ldots, m_k$, it is also divisible by their distinguished common multiple, that is,

$$[m_1, m_2, \ldots, m_k] \mid x_1 - x_2,$$

or written in congruence form,

$$x_1 \equiv x_2 \pmod{[m_1, m_2, \ldots, m_k]}. \tag{5}$$

We now assume that x_1 is a solution of (3) and additionally that (5) holds. We therefore have that

$$a_i x_1 \equiv b_i \pmod{m_i}, \qquad i = 1, 2, \ldots, k.$$

Since the m_i's are divisors of the distinguished common multiple, then by property (o) and (5), we have the following congruences:

$$x_1 \equiv x_2 \pmod{m_i}, \quad i = 1, 2, \ldots, k,$$

from which it follows that

$$a_i x_1 \equiv a_i x_2 \pmod{m_i}, \qquad i = 1, 2, \ldots, k.$$

The left-hand side is congruent to b_i modulo m_i, and therefore x_2 is also a solution of the congruences (3).

15. In what follows we will restrict ourselves to the special case mentioned above where the moduli are pairwise relatively prime. We will see that in this case the simultaneous congruence system always has a solution. We will give two methods to find it.

One of the methods goes through the individual congruences one by one, solving each based on the previous solution, ending with a solution for the entire system. The solution of the first congruence is a residue class $(c_1)_{m_1}$, i.e., the numbers of the form

$$x = c_1 + m_1 y_2,$$

where y_2 is an arbitrary integer.

Substituting this into the second congruence and rearranging, we have

$$a_2 m_1 y_2 \equiv b_2 - a_1 c_1 \pmod{m_2}.$$

This congruence has a solution, since by hypothesis, a_2 and m_1 are both relatively prime to m_2, and therefore so is their product. The solution of this is a residue class $(c_2)_{m_2}$, so we have

$$y_2 = c_2 + m_2 y_3,$$

where y_3 is an arbitrary integer. Substituting back for x, we get

$$x = c_1 + m_1 c_2 + m_1 m_2 y_3.$$

Continuing in this manner, we get a solution of the form

$$x = c_1 + m_1 c_2 + m_1 m_2 c_3 + \cdots + m_1 m_2 \cdots m_{k-1} c_k + m_1 m_2 \cdots m_k y_{k+1}, \quad (6)$$

where y_{k+1} is an arbitrary integer. We find solutions at each step (namely a residue class $(c_i)_{m_i}$) that are solutions to all previous congruences as well.

We need to show that this method will work at every step, in other words, that there is a solution at each step. If $2 \leq i \leq k$ and we have already determined $c_1, c_2, \ldots, c_{i-1}$, then by our method, the ith congruence will be rewritten

$$a_i(c_1 + m_1 c_2 + \cdots + m_1 m_2 \cdots m_{i-2} c_{i-1} + m_1 m_2 \cdots m_{i-1} y_i) \equiv b_i \pmod{m_i}.$$

Multiplying out, moving all but the last term to the right-hand side, and denoting the resulting value by d_i, we have

$$a_i m_1 m_2 \cdots m_{i-1} y_i \equiv d_i \pmod{m_i}.$$

This has a solution, since all of the factors on the left are relatively prime to m_i, hence so is their product. This proves our claim.

16. The method we just discussed depends on the order of the congruences; as a matter of fact, we can change this order to our advantage.

Notice that the next-to-last term of (6) plays a role only in the kth congruence. In all of the others it is congruent to 0. In fact, for every modulus we can find a term that is congruent to 0 with respect to every other modulus. In this way we can break up the unknown into subunknowns, each of which can be independently determined by one of the congruences.

To make things shorter, let M be the product of the moduli, and denote the quotient M/m_i by M_i. Write the unknown x as

$$x = \sum_{j=1}^{k} M_j z_j. \tag{7}$$

Substituting this into the ith congruence, we see that all terms on the left are congruent to 0 with the exception of the term $j = i$, since m_i is a factor of M_j for every $j \neq i$. We get the following congruences:

$$a_i M_i z_i \equiv b_i \pmod{m_i} \qquad i = 1, 2, \ldots, k.$$

The coefficient of z_i is relatively prime to the modulus, since a_i was assumed to be relatively prime to m_i, and M_i is the product of all moduli different from m_i, and these were also assumed to be relatively prime to m_i. These congruences therefore have solutions, and we have a new method for finding solutions of a system of simultaneous congruences in the special case considered. This is called the *Chinese remainder theorem*.

The Chinese remainder theorem gives a way to write a formula for the solution using additional quantities that are independent of the b_i's. Let a_i' and M_i' be solutions of the following congruences, respectively:

$$a_i a_i' \equiv 1 \pmod{m_i}, \quad M_i M_i' \equiv 1 \pmod{m_i}, \quad i = 1, 2, \ldots, k. \tag{8}$$

These have solutions by the fact that a_i, M_i are relatively prime to m_i. We then have a general solution of the form

$$x \equiv \sum_{i=1}^{k} a_i' M_i' M_i b_i \pmod{M}. \tag{9}$$

We summarize these results in the following theorem:

Theorem 7 *If the moduli in the simultaneous congruence system*

$$a_i x \equiv b_i \pmod{m_i}, \quad i = 1, 2, \ldots, k,$$

are pairwise relatively prime and

$$(a_i, m_i) = 1, \quad i = 1, 2, \ldots, k,$$

then the system has a solution. This solution is a residue class modulo the product of the moduli. The solution can be determined in the form of congruence (6), (7), *or* (9), *where additional quantities are defined by the congruences* (8).

17. In the previous section we stressed the advantages of the Chinese remainder theorem over the other method discussed. We note, however, that the Chinese remainder theorem does not help in the general case, while the other method does, by substituting the least common multiple of the moduli for their product.

If we do not have a restriction on the moduli of the congruence system (3), but instead we assume that the system has a solution, then the following method leads to a solution. Let

$$T_0 = 1, \quad T_1 = m_1, \quad T_i = [m_1, m_2, \dots, m_i], \quad i = 2, 3, \dots, k - 1. \quad (10)$$

Then we can find a solution of the form

$$x = \sum_{i=0}^{k-1} T_i u_i. \quad (11)$$

If at any step we get a congruence that does not have a solution, then the system (3) does not have a solution.

It is easy to see that a solution exists if and only if the following congruences are satisfied:

$$a_i b_j \equiv a_j b_i \pmod{(m_i, m_j)}, \qquad i, j = 1, 2, \dots, k. \quad (12)$$

(See Exercises 17 and 18.)

Exercises:

14. Show that any number satisfying congruence (9) is a solution of the simultaneous system of congruences (3).

15. Find a simple method to solve the following congruence system:

$$2x \equiv 5 \pmod{27}, \quad 18x \equiv 81 \pmod{69}, \quad 8x \equiv 12 \pmod{29}.$$

16. Twice a four-digit number has a remainder 1 upon division by 7, three times the number has a remainder 2 upon division by 8, and four times the number has a remainder 3 upon division by 9. What can this number be?

17. Prove that if system (3) has a solution, then the conditions (12) are satisfied.

18. Assume that congruence system (3) satisfies conditions (4) and (12). Let us introduce the notation

$$z_j = \sum_{i=0}^{j-1} T_i u_i, \qquad j = 1, 2, \ldots, k,$$

where the T_i are defined in (10). Prove that

(a) it is possible to choose z_1 so that it satisfies the first congruence; i.e., $a_1 z_1 \equiv b_1 \pmod{m_1}$;

(b) if z_{i-1} satisfies the first $i-1$ congruences with integers $u_0, u_1, \ldots, u_{i-2}$, then substituting z_i for x in the ith congruence, there is a solution for u_{i-1} (and therefore we have shown that the congruence system is solvable).

18. In this section and the next we mention two applications. The first is a faster algorithm for finding the result of a computation by using computers capable of running processes in parallel. Suppose we have a computer that can run k processes in parallel. We choose k primes $p_1, p_2, \ldots, p_k$ such that their product is larger than the expected solution. We divide the starting values by the p_i's, and we proceed with the computations in parallel using the remainders of the corresponding p_i moduli. Finally, we use the partial results to find a number having the desired residues for each of the given moduli, i.e., solving the simultaneous congruence system to get a solution for the congruence. The last step is very time costly, but we still have an improvement on the computing time.

19. We have already seen that arbitrarily large gaps can occur between neighboring primes (see Exercise 1.20*). As a second application we will show that there are *reclusive* primes, primes that are far from all other primes. More precisely, we will prove the following:

Theorem 8 *For any given number N, there exists a prime number that is at least N greater than the previous prime number and at least N smaller than the following one.*

For the proof of this theorem, we will use without proof the deep theorem of P. G. LEJEUNE DIRICHLET concerning prime numbers in arithmetic progressions.[8]

Dirichlet's Theorem. *If a and m are integers, relatively prime to each other, then there are infinitely many positive integers k such that $a + km$ is prime.*

---

[8] For a proof of Dirichlet's theorem we refer the reader to H. RADEMACHER: *Lectures on Elementary Number Theory*, Blaisdell Publ. Co., New York 1964. Cf. Chapter 14, pp. 121–136.

We note that the hypothesis of the theorem is necessary for there to be infinitely many primes in the sequence (why?).

We will base the proof on the idea that we choose $2N$ primes, $p_1, p_2, \ldots, p_{2N}$, and we look for a prime p such that for $i = 1, 2, \ldots, N, p - i$ is divisible by p_i; and $p + i$ is divisible by p_{N+i}. For p_i we will choose the ith prime.

We first try to find such a number p, ignoring the requirement that it should be prime. The number is a solution of the following simultaneous congruence system:

$$x \equiv j \pmod{p_j}, \quad x \equiv -j \pmod{p_{N+j}}, \quad j = 1, 2, \ldots, N.$$

Since the moduli are pairwise relatively prime, we know that the system has a solution, namely a residue class

$$x \equiv x_0 \pmod{p_1 p_2 \cdots p_{2N}}.$$

So the solutions are the numbers of the form

$$x_0 + k p_1 p_2 \cdots p_{2N},$$

where k is any integer. This is an arithmetic progression, and by Dirichlet's theorem it contains infinitely many primes as long as $(x_0, p_1 p_2 \cdots p_{2N}) = 1$. This condition is immediate, because if the greatest common divisor were greater than one, it would have to be divisible by one of the p_i's. This would mean on the one hand that x_0 is divisible by p_i, and on the other hand, one of the two congruences for p_i above holds, depending on whether $i \leq N$ or $i > N$. In both cases the congruence implies that $p_i | j$, which is impossible, since $p_i > j$.

Hence there are infinitely many primes of the form $x_0 + k p_1 p_2 \cdots p_{2N}$, where k is a positive integer. For primes p of this form, $p + i$ is composite, since it is divisible by p_i and greater than it. Similarly, $p - i$ is also composite because it is divisible by p_{N+i}, and from $p_1 = 2$ it follows that

$$p - i = x_0 + k p_1 p_2 \cdots p_{2N} - i > 2 p_{2N} - i > p_{2N} + 2N - i > p_i.$$

This finishes the proof.

Exercises:

19. Give a proof of Theorem 8 analogous to EUCLID's proof of the existence of infinitely many primes, showing that there is a prime p for a suitable prime q, such that $p + i$ is divisible by $q + i$ and $p - i$ is divisible by $q - i$, for $i = 1, 2, \ldots, q - 2$.

20. Prove that for an arbitrary positive integer N:
    (a) There exist N consecutive integers, each divisible by a square greater than 1.
    (b) There exist N consecutive integers, none of which is a power of an integer (where the exponent is greater than one).

20. We have seen that every element of a residue class has the same g.c.d. with the modulus, and we mentioned that those classes that are relatively prime to the modulus (i.e., those made up of numbers relatively prime to the modulus) play an important role. For example, precisely these classes have reciprocals. Indeed, if $(a, m) = 1$, then there is a number a' for which $(a)_m(a')_m = (1)_m$, i.e.,

$$aa' \equiv 1 \pmod{m},$$

which we have already seen in the discussion of the Chinese remainder theorem. For other residue classes, this statement is not true.

A set of representatives, one from each class relatively prime to the modulus, is called a *reduced residue system*. It is customary to denote by $\varphi(m)$ the number of residue classes relatively prime to m. This is called *Euler's φ-function* after its introducer. Its definition, of course, does not require congruences. In a complete residue system, one representative appears from each class, and $\varphi(m)$ is the number of those representatives relatively prime to the modulus. By choosing a special residue system, namely the smallest nonnegative representative from each class, we arrive at the following definition: $\varphi(m)$ *is the number of integers from 0 to $m - 1$ that are relatively prime to m.* For example, $\varphi(1) = \varphi(2) = 1$, $\varphi(3) = \varphi(4) = 2, \varphi(5) = 4, \varphi(6) = 2,$ $\varphi(7) = 6$. In general, if p is a prime, then among the numbers $0, 1, \ldots, p - 1$, only the first is not relatively prime to p; hence

$$\varphi(p) = p - 1.$$

21. Euler's φ-function has an interesting and surprisingly easily provable property, namely that for a given n, if we sum the values of the φ-function for all divisors of n, we get precisely n. We will denote the summation over all divisors d of the number n by $\sum_{d\mid n}$ and formulate the property in the following theorem:

Theorem 9 *For Euler's φ-function,*

$$\sum_{d\mid n} \varphi(d) = n.$$

Having this relation we can even forget the original definition, and it is still possible to calculate the value of the function for every integer. This relation gives, namely, for $n = 1$,

$$\varphi(1) = \sum_{d\mid 1} \varphi(d) = 1;$$

if $n > 1$, we can rewrite the above as

$$\varphi(n) = n - \sum_{d|n}^{*} \varphi(d),$$

where the star in the summation means that the sum excludes $d = n$. From this it follows that

$$\varphi(2) = 2 - \varphi(1) = 1, \quad \varphi(3) = 3 - \varphi(1) = 2, \quad \varphi(4) = 4 - \varphi(2) - \varphi(1) = 2,$$

and we can similarly calculate the value of the function for every integer if we know its value for the smaller integers. Such a way of describing a function is called a *recursive definition*.

For the proof of the theorem, reduce the following fractions to their simplest form:

$$\frac{1}{n}, \quad \frac{2}{n}, \quad \ldots, \quad \frac{n-1}{n}, \quad \frac{n}{n}.$$

It is clear that after simplifying, all denominators are divisors of n and the numerators are relatively prime to their denominators; all fractions of this type not greater than 1 appear, and only these. For a divisor d of n, there are exactly $\varphi(d)$ such fractions with denominator d, thus proving the theorem.

We will later give a more effective method for calculating the function, as well as a useful application of the theorem.

As an application whose result we will need shortly, let us determine $\varphi(pq)$ for distinct primes p and q. By the recursion, we have

$$\varphi(pq) = pq - \varphi(p) - \varphi(q) - 1 = pq - (p - 1) - (q - 1) - 1 = (p - 1)(q - 1).$$

22. For the proof of an important property of the φ-function, we start with the following analogue of Theorem 1:

Theorem 10 *If $\varphi(m)$ integers are relatively prime to m and are pairwise incongruent modulo m, then they form a reduced residue system modulo m.*

It is clear that reduced residue systems have these properties. Assume that we have a system satisfying the hypotheses of the theorem. Since the elements are relatively prime to m, they are elements of residue classes relatively prime to m. Because they are pairwise incongruent modulo m, they come from distinct residue classes. Finally, there are $\varphi(m)$ of them, so there must be one representative from each residue class relatively prime to m. Thus this is a reduced residue system, as stated.

Exercise:

21. (a) Find the value of $\varphi(p^m)$ where p is a prime number and m is any positive integer.

(b) Prove that for positive integers a, b, if $(a, b) = 1$, then $\varphi(ab) = \varphi(a)\varphi(b)$.

(c) Give a formula for calculating the value of the φ-function.

23. We now come to the important property mentioned above. Let m be an integer greater than 1, a an integer relatively prime to m, and $r_1, r_2, \ldots, r_{\varphi(m)}$ a reduced residue system modulo m. Then according to Theorem 10,

$$ar_1, ar_2, \ldots, ar_{\varphi(m)}$$

is a reduced residue system as well, because it has $\varphi(m)$ elements that are relatively prime to m (since each is a product of two integers relatively prime to m) and these elements are pairwise incongruent modulo m. In fact, if we assume that

$$ar_i \equiv ar_j \pmod{m},$$

then by property (i) and the fact that $(a, m) = 1$, we can simplify to get

$$r_i \equiv r_j \pmod{m},$$

and in a reduced residue system this is possible only if $i = j$.

For each element in the new reduced residue system there is exactly one element congruent to it from the original system:

$$ar_i \equiv r_{j_i} \pmod{m}, \quad i = 1, 2, \ldots, \varphi(m).$$

Here $j_1, j_2, \ldots, j_{\varphi(m)}$ are the numbers $1, 2, \ldots, \varphi(m)$, generally in some other order.

In the above congruences, multiply all the left sides together and all the right sides together. In this way, we get

$$a^{\varphi(m)} r_1 r_2 \cdots r_{\varphi(m)} \equiv r_{j_1} r_{j_2} \cdots r_{j_{\varphi(m)}} = r_1 r_2 \cdots r_{\varphi(m)} \pmod{m}.$$

The product of the r's appearing on both sides is made up of factors relatively prime to m, and hence the product itself is relatively prime to m. By property (i) we may divide by this product to get the following result:

Theorem 11 (Euler–Fermat Theorem) *If m is an integer greater than 1 and a is an integer relatively prime to m, then*

$$a^{\varphi(m)} \equiv 1 \pmod{m}.$$

24. In the case where the modulus is a prime p, we have $\varphi(p) = p - 1$, and $(a, p) = 1$ means that $p \nmid a$. Hence we get the following result:

Theorem 11′ (Fermat's Theorem) *If p is a prime and $p \nmid a$, then*

$$a^{p-1} \equiv 1 \pmod{p}.$$

This is due to FERMAT, but unfortunately we do not have his proof. The first proofs we have of this theorem and its generalization, the Euler–Fermat theorem, are due to EULER. The proof given here is essentially due to him as well.

In the prime modulus case, we can drop the condition that $p \nmid a$ by rephrasing Fermat's theorem in the following equivalent form:

Theorem 11″ *If p is a prime, then for every integer a,*

$$a^p \equiv a \pmod{p}. \tag{13}$$

If $p \mid a$, then both sides of the congruence are congruent to 0; otherwise, Theorem 11′ holds, and multiplying both sides by a we get (13).

We still have to prove that Theorem 11′ follows from this theorem. Assume that (13) holds and $p \nmid a$. Then $(a, p) = 1$, and by property (i) we can divide both sides of the congruence by a to get Fermat's theorem.

25. The Euler–Fermat theorem provides a new method to solve the congruence $ax \equiv b \pmod{m}$ in the case where $(a, m) = 1$. According to Theorem 4, the congruence is solvable, and the solution is one residue class modulo m. We can solve the congruence by multiplying both sides by $a^{\varphi(m)-1}$. The new coefficient of x is congruent to 1, and we get

$$x \equiv a^{\varphi(m)-1}b \pmod{m}.$$

The solution of the congruence is therefore the residue class

$$\left(a^{\varphi(m)-1}b\right)_m. \tag{14}$$

26. Though it is customary to call this last theorem Fermat's *little* theorem, we will see that it plays an important role, so we give another proof of Theorem 11″ based on an idea completely different from that above.

It is enough to prove the theorem for odd primes, since it is obviously true for $p = 2$. Let c be a positive integer and number the vertices of a regular p-gon in all possible ways using the numbers $1, 2, \ldots, c$. In this way we get c^p labeled polygons.

Take an arbitrary labeled polygon C_0 and rotate it repeatedly by $360/p$ degrees about its center. In this way we get a sequence of polygons $C_0, C_1, C_2, \ldots$. Obviously, $C_0 = C_p$, and the sequence is periodic. Let C_i be the first polygon that is the same as an earlier C_j. Then $j = 0$, for otherwise C_{i-1} would be the same as C_{j-1}. From the ith on, the first i pairwise different polygons recur periodically . Those polygons C_k for which k is a multiple of i are the same as C_0. The polygon C_p is among these, from which we see that $i \mid p$.

Since p is prime, either $i = 1$ or $i = p$. If $i = 1$, then this means that all vertices of the polygon have the same label. The number of these polygons is c. We can put the remaining polygons into classes, in such a way that two polygons are put in the same class if one can be rotated into the other. All classes will have size p. In this way, we see that for any positive integer $c, p \mid c^p - c$, and hence $c^p \equiv c \pmod{p}$.

Finally, if a is any integer, then there is a positive integer c for which $a \equiv c \pmod{p}$, and by property (m) we have

$$a^p \equiv c^p \equiv c \equiv a \pmod{p}.$$

This concludes the new proof of Theorem 11″ for any integer a.

27. Since the 1970s, number theory has played an important role in cryptography. The basis of cryptography is encoding messages. For instance, the sender S (Susan) wants to send message m to the addressee A (Adam). First she encodes the message m by using function f, getting $M = f(m)$, the encoded message. For example, she can replace every letter by a previously agreed-upon letter (replace a by c, b by q, etc.); or all letters could be shifted by three characters in the alphabet; or every character could be replaced by a two-digit number; or any of a number of other clever methods.

Then S sends M to A. If A knows the inverse of f, f^{-1}, then he can determine the original message,

$$m = f^{-1}(M) = f^{-1}(f(m)). \tag{15}$$

In order for them to correspond securely, both must know f and f^{-1} and they must keep these secret.

In the above examples it was easy to determine M from m and vice-versa. We could also have used f^{-1} for encoding the message, and then f is used to decode it:

$$f(f^{-1}(m)) = f^{-1}(f(m)) = m. \tag{16}$$

Those people who intercept encoded messages between S and A try to determine f and f^{-1} and hence the original messages. To make it more difficult for an outsider to break the code, it is advisable for S and A to change the encoding and decoding methods from time to time, of course both changing at the same time. But this further complicates communication.

If it were possible to find an encoding scheme for which it is impossible to find f^{-1}, even knowing f, then S could publish her own f_S, whose inverse f_S^{-1} only she knows. Then anyone can send her a message that only she can decode.

Is such a scheme possible? The first impression is that the answer is no, but with the help of the Euler–Fermat theorem, and the help of the processing power of computers on the one hand and their inherent limitations on the other, it is possible.

28. We already mentioned that messages can be converted to numbers, and from here we will deal only with encoding of numbers. Choose a number N that is the product of two distinct primes p and q. Break the number we want to send into parts smaller than N (for instance, if $N = 4757(= 67 \cdot 71)$ and we want to send the number $m = 1234567890$, a possible break-up is 1234, 567, 890). Let one of the parts be h, so that

$$0 < h < N. \tag{17}$$

We further choose two numbers r and s such that $rs \equiv 1 \pmod{\varphi(N)}$, so that for some integer $k, rs = 1 + k\varphi(N)$. We have seen in Section 21 that $\varphi(N) = (p-1)(q-1)$. Then the encoding and decoding are done in the following way:

- Let $H = f(h)$ be the smallest nonnegative remainder of h^r, modulo N.
- Let $f^{-1}(H)$ be the smallest nonnegative remainder of H^s, modulo N.

We will show that these f and f^{-1} are indeed inverses of each other, i.e., satisfy (16). Now

$$f^{-1}(H) = f^{-1}(f(h)) \equiv h^{rs} = h^{1+k\varphi(N)} = hh^{k\varphi(N)} \pmod{N}.$$

If $(h, N) = 1$, then according to the Euler–Fermat theorem

$$h^{k\varphi(N)} = (h^{\varphi(N)})^k \equiv 1 \pmod{N}; \quad \text{hence} \quad h^{rs} \equiv h \pmod{N},$$

and by (17), the right-hand side is the smallest nonnegative remainder of the left-hand side, modulo N, thus satisfying (16).

If $(h, N) = p$, then we may write $h = p^u h'$ where $pq \nmid h'$, i.e., $(h', N) = 1$. By Fermat's theorem and property (m),

$$h^{q-1} \equiv 1 \pmod{q}; \quad \text{hence} \quad h^{k(p-1)(q-1)} \equiv 1 \pmod{q}.$$

Multiplying both sides by $p^{u-1}h'$ and then both sides and the modulus by p, we have then multiplied both sides by h and the modulus by p. Using the facts that $pq = N$ and $(p-1)(q-1) = \varphi(N)$, we see that

$$h^{rs} = hh^{k\varphi(N)} \equiv h \pmod{N}.$$

It is obvious that the congruence $h^{rs} \equiv h \pmod{N}$ is valid even in the case where $N \mid h$ (notice, however, that this was excluded by our scheme).

29. With this we have found an encoding scheme, and by the commutative property of multiplication, (16) is also satisfied. The question only remains as to why we can publish f, which means making N and r known? Here the computer comes into play. First we need two prime numbers, very large for two reasons. The first so that we do not need to break the message up into too many parts. The second, more important, we will soon see.

Today, there are algorithms that when implemented on a powerful enough computer can decide in a short amount of time with an incredibly small possibility of error whether or not a 100-digit number, or even larger, is prime.

We can cleverly choose large numbers that look suspiciously like primes. We then test whether they are prime, and in this way we can find large enough primes p and q. We calculate N and $\varphi(N)$ from these. Then we choose an r, checking its relative primeness with $\varphi(N)$ using the Euclidean algorithm, and finally determine s by solving the congruence $rs \equiv 1 \pmod{\varphi(N)}$, again using the Euclidean algorithm.

We can now publish N and r, keeping p, q, and s secret. We need not worry about publishing these, since in order for someone to find our code, they "only" need to factor N. To do this, they "only" need to check whether N is divisible by any of the primes up to $\sqrt{N}$. It is just a small part of the problem that we do not know which are prime, because we can check all odd numbers up to this bound, excluding those numbers that are divisible by 3 and 5, since these are easily identifiable. The set of integers to be checked can be further reduced by some clever tricks, but this is only a minimal gain over checking all numbers; it helps when we are checking numbers up to a million or a billion, but when there are hundreds of digits in the number, this gain is insignificant, and the process of dividing every number can take millions of years even on today's fastest supercomputers. The rate of increase in processing power in computers and the physical limits for computing speed are such that we cannot hope to decrease this computing time significantly. It also seems very unlikely that any major mathematical advances will be made for simplifying this factorization. So at least at the present time, everyone can publish his or her own N and r without worrying, just keeping p, q, and s under lock and key.

30. There is the danger in making the code (N and r) known that a person can send phony messages as well as forge messages in someone else's name, in this way endangering the addressee and/or the sender. However, (16) can serve as additional protection against this type of attack. Susan, after encoding the message m using Adam's public key, can encode the message $f_A(m)$ using her private key, obtaining the message $f_S^{-1}(f_A(m))$, which she sends to Adam. This message could only have been produced by Susan and can only be decoded by Adam. Adam decodes this by first applying f_S and then f_A^{-1}, getting

$$f_A^{-1}(f_S(f_S^{-1}(f_A(m)))) = m.$$

In this way, it is possible to sign contracts, transfer money, and share information without traveling across continents.

These discoveries have opened new chapters in cryptography, and there are already large amounts of literature written in this area. Unfortunately, these techniques are offered not only in the financial and diplomatic worlds, but for organized crime as well, creating serious problems.

31. We now return to the Euler–Fermat theorem. According to this theorem, for an arbitrary modulus m and every number relatively prime to the modulus, there exists a power of this number that is congruent to 1 modulo m, for instance the $\varphi(m)$th power. The smallest such exponent could be smaller than $\varphi(m)$. For example, 1 raised to the first power is already congruent to 1 for every modulus, and the second power of $m - 1$ is also congruent to 1, modulo m.

We call the smallest exponent of a number c for which c raised to this exponent is congruent to 1 modulo m the *order of c modulo m*. Only those numbers that are relatively prime to the modulus can have an order, and by the Euler–Fermat theorem, they do also have one.

32. The numbers, or more precisely the residue classes, that have order k modulo m satisfy the congruence $x^k - 1 \equiv 0 \pmod{m}$, and hence we should become more familiar with polynomials in congruences. Only after this little detour will we return to orders of elements.

By property (n), the number of solutions here is understood as the number of residue classes. From algebra, we know that the number of roots of an algebraic equation, i.e., an equation of the form $f(x) = 0$, where f is a polynomial, cannot be larger than the degree of the polynomial. The congruence

$$x^2 - 1 \equiv 0 \pmod{24},$$

however, has solutions 1, 5, 7, 11, 13, 17, 19, 23, which are all representatives of distinct residue classes.

We also note that the coefficients of a polynomial are only representatives of their residue class. In this way, we see that the polynomial $35x^5 - 21x^4 + 23x^2 + 14x - 7$ is a fifth-degree polynomial modulo most primes, but modulo 5 it is of degree 4, and modulo 7 the polynomial reduces to the term $23x^2$, which is also equivalent to $2x^2$, hence of degree 2. For this reason, we consider two polynomials the same if their corresponding coefficients are congruent.

The proof of the previously mentioned algebraic theorem uses the fact that a factor corresponding to a root can be factored out. This also holds for polynomials in congruences.

Theorem 12 *If the coefficients of a polynomial $f(x)$ are not all congruent to 0 modulo m and $f(a) \equiv 0 \pmod{m}$, then there exists a polynomial $g(x)$ such that*

$$f(x) \equiv (x - a)g(x) \pmod{m}$$

for all x and the degree of g is one less than the degree of f.

Let $f(x)$ be a polynomial of degree n modulo m and suppose that $f(a) \equiv 0 \pmod{m}$. Consider the polynomial $F(x) = f(x + a)$, whose constant term is c_0. We can write this

$$F(x) = c_0 + xG(x),$$

where $G(x)$ is of degree $n - 1$ (it is not hard to see that the coefficient of the term of highest degree is the same as for f). Therefore,

$$F(0) = c_0 = f(a) \equiv 0 \pmod{m}.$$

Returning to f, we have

$$f(x) = F(x - a) = c_0 + (x - a)G(x - a) \equiv (x - a)G(x - a) \pmod{m}.$$

Hence letting $g(x) = G(x - a)$, we have found such a polynomial g satisfying the claim.

33. The proof of the previously mentioned theorem regarding the number of roots of a polynomial also uses the fact that a product is 0 only if one of its factors is 0. Among congruences, this is in general not true. Returning to the example above, the polynomial $x^2 - 1$ has $x - 7$ as one of its factors, modulo 24, and may be factored

$$x^2 - 1 \equiv (x - 7)(x - 17) \pmod{24}.$$

Substituting 11 in the right-hand side we get the congruence

$$4 \cdot (-6) \equiv 0 \pmod{24},$$

where neither of the factors of the product is 0 (they are not divisible by 24).

This statement regarding the vanishing of a product is, however, valid for prime moduli. This is just the prime property, and in this way we get the following:

Theorem 13 *For a prime modulus, a nonzero polynomial cannot have more (residue classes as) roots than the degree of the polynomial.*

We prove this theorem using induction on the degree of the polynomial. For degree one polynomials the theorem is true by Theorem 4.

Let p be a prime, k an integer greater than 1, and assume that the theorem is true for polynomials of degree $k - 1$. If $f(x)$ is a polynomial of degree k and

$$f(x) \equiv 0 \pmod{p} \tag{18}$$

has no solution for an integer x, or the elements of only one residue class satisfy the congruence, then the theorem is true for f as well. If the congruence holds for a and b (from different residue classes) modulo p, then by Theorem 12 there is a polynomial g of degree $k - 1$ for which

$$f(x) \equiv (x - a)g(x) \pmod{p}.$$

Substituting in b, we have

$$(b - a)g(b) \equiv 0 \pmod{p}.$$

The first factor here is not congruent to 0, i.e., not divisible by the prime p, and therefore by the prime property, the second factor is divisible by p, meaning that it is congruent to 0 modulo p. Therefore, every solution of (18) not congruent to a modulo p is a solution of

$$g(x) \equiv 0 \pmod{p}.$$

The polynomial g is of degree $k - 1$, and by induction it has at most $k - 1$ roots, so f can have at most k roots. This concludes the proof of the theorem.

34. As an application of the above theorem, we prove the following theorem, due to WILSON:

Theorem 14 (Wilson's Theorem) *If p is a prime number, then*

$$(p-1)! + 1 \equiv 0 \pmod{p}.$$

This characterizes prime numbers, since no composite number will ever satisfy the above congruence (see Exercise 26).

Using Theorem 13 we can find a simple proof for the theorem. (Later, when we discuss second-degree congruences, we will see a different proof of this theorem, and in Chapter 4 we will give a geometric proof.)

For $p = 2$, the statement is clearly true, and we may assume p to be an odd prime. The degree of the polynomial

$$(x-1)(x-2)\cdots(x-(p-1)) - \left(x^{p-1} - 1\right) = a_{p-2}x^{p-2} + a_{p-3}x^{p-3} + \cdots + a_0,$$

if it exists (if it is not the zero polynomial), is at most $p - 2$, since the term of degree $p - 1$ cancels out. However, this polynomial has $p - 1$ distinct roots, since the first term $(x-1)(x-2)\cdots(x-(p-1))$ is 0 for all numbers $1, 2, \ldots, p-1$, and the second term $\left(x^{p-1} - 1\right)$ is congruent to 0 modulo p for these numbers by Fermat's theorem (Theorem 11′). Therefore, their difference is congruent to 0 modulo p. This is possible only if the polynomial on the right-hand side is the zero polynomial, in which case the remaining residue class $(0)_p$ is also a root. Substituting this into the congruence, we get

$$a_0 = (-1)^{p-1}(p-1)! - (-1) = (p-1)! + 1 \equiv 0 \pmod{p},$$

proving the claim of the theorem.

We could have also concluded the proof in the following way: The conclusion that the polynomial is the zero polynomial means that every coefficient must be congruent to 0 modulo p,

$$a_{p-2} \equiv a_{p-3} \equiv \cdots \equiv a_1 \equiv a_0 \equiv 0 \pmod{p}.$$

Thus the proof furnishes a series of congruences; the last one is Wilson's theorem, having indeed the most applications.

35. From Wilson's theorem we can deduce the following theorem, which we will make use of later in the book.

Theorem 15 *For all primes p of the form $4k + 1$, there exists an integer c for which*

$$c^2 + 1 \equiv 0 \pmod{p}.$$

Using the remark that

$$p - j \equiv -j \pmod{p}, \qquad j = \frac{p-1}{2}, \frac{p-1}{2} - 1, \ldots, 2, 1,$$

Wilson's theorem can be rewritten as follows:

$$(-1)^{\frac{p-1}{2}}\left(\left(\frac{p-1}{2}\right)!\right)^2 + 1 \equiv 0 \pmod{p}.$$

If p is of the form $4k + 1$, then -1 is raised to an even power, and thus $c = ((p-1)/2)!$ satisfies the claim of the theorem.

Exercises:

22. For a composite number m, determine what $(m-1)! + 1$ is congruent to modulo m.

23. Prove that numbers of the form $c^2 + 1$ are not divisible by any primes of the form $4k + 3$. (Cf. Theorem 1.16.)

24*. Using the notation of the previous section, prove that if $p \geq 5$, then a_1 is also divisible by p^2 (Wolstenholme's theorem).

25*. Prove that if p is an odd prime greater than $3, k$ is an integer $2 \leq k \leq p - 2$, and $r_1, r_2, \ldots, r_{p-1}$ forms a reduced residue system modulo p, then
    (a) there exists an integer r for which $r \not\equiv 0, r^k \not\equiv 1 \pmod{p}$
    (b) $r_1^k + r_2^k + \cdots + r_{p-1}^k \equiv 0 \pmod{p}$.

26. For which integers m and $k, k > 1$, is the set $0, 1, 2^k, \ldots, (m-1)^k$ a complete residue system modulo m?

36. At the beginning of Section 32 we mentioned that we need to investigate the number of roots of a polynomial of the form $x^k - 1$, modulo m. With this in mind, we have the following result.

Theorem 16 *If p is an odd prime and $k \mid p - 1$, then the congruence*

$$x^k - 1 \equiv 0 \pmod{p}$$

has k roots.

It is well known that if $k \mid (p-1)$, then the polynomial $x^{p-1} - 1$ factors into

$$x^{p-1} - 1 = \left(x^k - 1\right) g(x),$$

where $g(x)$ is some polynomial of degree $p - 1 - k$. By Fermat's theorem, the left-hand side has $p - 1$ roots modulo p. Since we are dealing with a prime modulus, the right-hand side can be congruent to 0 modulo p only if one of its factors is congruent to 0. The first factor has at most k roots and the second at most $p - 1 - k$ roots. If one of these factors had fewer than this many, then their product would have fewer than $p - 1$ roots. Therefore, both factors have their maximum number of roots, proving the theorem.

Exercises:

27. Prove that if $(k, p - 1) = d$, then the congruence

$$x^k - 1 \equiv 0 \pmod{p}$$

has d roots.

28. Prove that
  (a) if the congruence

$$x^k \equiv c \pmod{p} \tag{19}$$

is solvable, then the congruence

$$c^{\frac{p-1}{(k,p-1)}} \equiv 1 \pmod{p} \tag{20}$$

holds, and (19) has $(k, p - 1)$ residue classes as solutions.
  (b) If (20) is satisfied, then (19) has a solution.

37. We now return to the topic of the order of elements. In the following table we list the minimal exponent for all numbers relatively prime to the given moduli:

modulus:	6		9						15							
number:	1	5	1	2	4	5	7	8	1	2	4	7	8	11	13	14
exponent:	1	**2**	1	6	3	6	3	2	1	4	2	4	4	2	4	2

The value of Euler's φ-function for these numbers is 2, 6, and 8, respectively. We used boldface for those exponents that equal the corresponding value of the φ-function. In the case of 15, there is no such exponent.

We also see in the above examples that the exponents are divisors of the corresponding values of φ. This is not by accident, and in fact even more is true.

Theorem 17 *If a number c has order n modulo m, then the numbers $1, c, c^2, \ldots, c^{n-1}$ are pairwise incongruent modulo m. If*

$$c^u \equiv c^v \pmod{m}, \qquad then \quad u \equiv v \pmod{n}.$$

In the special case where $v = 0$, i.e., $c^u \equiv 1 \pmod{m}$, then $n \mid u$.

According to the theorem, it then follows from the Euler–Fermat theorem that $n \mid \varphi(m)$.

The second claim of the theorem follows from the first, so it is enough to prove that. Without loss of generality, we may assume that $u \geq v$. Then $c^{u-v} \equiv 1 \pmod{m}$. Dividing $u - v$ by n with remainder we get

$$u - v = nq + s,$$

where q is an integer and $0 \le s < n$. Based on this we have

$$c^{u-v} = (c^n)^q c^s \equiv c^s \equiv 1 \pmod{m}.$$

Among the positive integral powers of c the nth is the first that is congruent to 1 modulo m. Therefore, $s = 0$, and hence $n \mid u - v$.

38. Those elements g whose order is $\varphi(m)$, if such elements exist, play an important role. The first $\varphi(m)$ powers of these, $g, g^2, \dots, g^{\varphi(m)}$, by Theorems 9 and 12, form a reduced residue system modulo m. These numbers that have order $\varphi(m)$ are called *primitive roots of congruence modulo* m, or just primitive roots. If for a given modulus m there exists a primitive root g, then for every c that is relatively prime to m, we call the smallest nonnegative exponent k for which

$$c \equiv g^k \pmod{p}$$

the *index (with respect to g) of c modulo m.*

We will show that for every prime p, there exists a primitive root modulo p. Further, we will determine all n such that there are elements of order n and how many such elements there are.

Theorem 18 *The number of elements of order n modulo a prime p is $\varphi(n)$ if $n \mid p-1$; otherwise, it is 0. Based on this, there are $\varphi(p-1)$ primitive roots modulo p.*

For an arbitrary $m > 0$, it follows by Theorem 17 that only divisors of $\varphi(m)$ can possibly occur as orders of elements. Thus the second claim is true.

We prove the first claim by induction on n. Let $\varrho(n)$ denote the number of residue classes of order n. If $n = 1$, the only element whose order is 1, is 1 itself, i.e., $\varrho(1) = 1 = \varphi(1)$, so the theorem is true for $n = 1$.

Let us now assume that $m > 1, m \mid p - 1$, and that the theorem is true for all $d < m$. The elements of order m (if any) are roots of the congruence

$$x^m - 1 \equiv 0 \pmod{p}.$$

On the one hand, this congruence has m roots by Theorem 16, and on the other hand, the other roots of this congruence (if any) are, by Theorem 17, those elements whose order d is a proper divisors of m. Using the induction hypothesis and Theorem 9 (as well as the notation introduced there), we see that the number of elements of order m is

$$m - \sum_{d \mid m}^{*} \varrho(d) = m - \sum_{d \mid m}^{*} \varphi(d) = \varphi(m).$$

Hence the theorem is also true for m, and we have proved the first statement of the theorem.

Exercises:

29. Without using the Euler–Fermat theorem, show that every number relatively prime to m has an order modulo m.

30. Given the indices of numbers a and b, what are the indices of a^h (for h a positive integer), ab, and the index of a number c for which $ac \equiv b$ (mod m)?

31. Show that
    (a) if the positive integer m has at least two distinct odd prime divisors, and c is relatively prime to m, then $c^{\varphi(m)/2} \equiv 1 \pmod{m}$;
    (b) if $k \geq 3$ and c is odd, then

    $$c^{2^{k-2}} \equiv 1 \pmod{2^k};$$

    (c) if the canonical decomposition of m is $m = \prod_{i=1}^{r} p_i^{a_i}$ and

    $$\beta(m) = \left[\varphi\left(p_1^{a_1}\right), \varphi\left(p_2^{a_2}\right), \ldots, \varphi\left(p_r^{a_r}\right)\right],$$

    then for every c relatively prime to m,

    $$c^{\beta(m)} \equiv 1 \pmod{m}.$$

The claims of the exercises show that there is no primitive root if the modulus is divisible by two odd primes, nor if the modulus is divisible by both 4 and a prime larger than 4. Thus there can be a primitive root only if the modulus is 2, 4, an odd prime power, or twice an odd prime power. It can be proved that for these moduli there indeed exist primitive roots. Theorem 18 gives them in the case of an odd prime modulus, and the cases of 2 and 4 are trivial.

39. We have already seen many interesting applications of Euler's φ-function, and later we will see more, so it is useful to investigate it even further. Theorem 9, for which we have already seen an important application, gives a method for calculating the value of the φ-function, but this method is rather uncomfortable to use. The results of Exercise 21 lead to an explicit formula based on a property that is called multiplicativity. A function f defined on the positive integers is called *multiplicative* if for all pairs of relatively prime integers a and b, it satisfies $f(ab) = f(a)f(b)$. If this is satisfied for all a and b, then the function is called *totally* or *completely multiplicative*. We will show in another way that Euler's φ-function is multiplicative.

Theorem 19 *Euler's φ-function is multiplicative.*

We need to show that if a and b are relatively prime, then

$$\varphi(ab) = \varphi(a)\varphi(b). \tag{21}$$

This is clearly true if either a or b is 1. If both are greater than 1, we need to investigate the numbers n between 0 and $ab - 1$ that are relatively prime to ab, in some way relating these to a and b. Consider the Diophantine equation

$$ax + by = n. \tag{22}$$

Since a and b are relatively prime, Theorem 1.18 guarantees that this equation has a solution for every n. For x and y satisfying (22), we have the following congruences:

$$ax \equiv n \pmod{b} \quad \text{and} \quad by \equiv n \pmod{a}. \tag{23}$$

By Theorem 4 these congruences have one residue class each as a solution. In this way, for every n we get a unique pair of residue classes modulo a and modulo b, respectively.

Conversely, consider now a pair of residue classes modulo b and modulo a, respectively. Choose an element x from the first and an element y from the second. Substituting into (23) and solving for n as the unknown in the simultaneous system of congruences, we see by Theorem 7 that there is a unique residue class modulo ab that is a solution, and this has a unique representative between 0 and $ab - 1$.

The important observation is that if n is relatively prime to ab, then x is relatively prime to b and y to a, and the converse is true as well. If a and y, or x and b, have a common divisor larger than 1, then this divisor would also divide n and ab, but we assumed that these are relatively prime. Conversely, if n and ab are not relatively prime, then they have a common prime divisor, and by the prime property this must divide either a or b. We may assume that it divides a. This divisor divides the left-hand side of (22) as well as the first summand. Therefore, it divides the product by. It does not divide b, because a and b are relatively prime. By Euclid's lemma it must divide y; hence a and y are not relatively prime. We have seen, therefore, that in the above construction n is relatively prime to ab if and only if x is relatively prime to b, and y to a.

Summarizing our results, we have paired up numbers between 0 and $ab-1$ that are relatively prime to ab with ordered pairs of numbers, the first between 0 and $b - 1$, relatively prime to b, and the second between 0 and $a - 1$ and relatively prime to a, thus proving (21) and hence the theorem.

By repeatedly using Theorem 19 for pairwise relatively prime positive integers $n_1, n_2, \ldots, n_k$, we have that

$$\varphi(n_1 n_2 \cdots n_k) = \varphi(n_1)\varphi(n_2) \cdots \varphi(n_k).$$

These conditions are satisfied for the prime power factors of the canonical decomposition of a number. If $n = p_1^{u_1} p_2^{u_2} \cdots p_k^{u_k}$ is the canonical decomposition of a number n, then we get the following equation:

$$\varphi(n) = \varphi\left(p_1^{u_1} p_2^{u_2} \cdots p_k^{u_k}\right) = \varphi\left(p_1^{u_1}\right) \varphi\left(p_2^{u_2}\right) \cdots \varphi\left(p_k^{u_k}\right).$$

For this formula to be useful, we have to calculate the φ-function for prime powers. We have seen that if p is a prime, then $\varphi(p) = p - 1$. If u is an integer greater than 1, then all the numbers $0, 1, 2, \ldots, p^u - 1$ that are not divisible by p are relatively prime to p^u. The numbers divisible by p are $0, p, 2p, \ldots, p^u - p$. There are p^{u-1} of these, so

$$\varphi\left(p^u\right) = p^u - p^{u-1} = p^{u-1}(p-1) = p^u \left(1 - \frac{1}{p}\right).$$

This formula includes the case $u = 1$ as well.

Putting these together, we get the following formula:

$$\varphi(n) = p_1^{u_1-1} p_2^{u_2-1} \cdots p_k^{u_k-1} (p_1 - 1)(p_2 - 1) \cdots (p_k - 1)$$
$$= n \left(1 - \frac{1}{p_1}\right)\left(1 - \frac{1}{p_2}\right) \cdots \left(1 - \frac{1}{p_k}\right).$$

40. The general questions concerning congruences of higher degree can be reduced essentially to investigating congruences for prime moduli. If a congruence is satisfied for a composite modulus m, then it is satisfied modulo all the prime divisors of m.

Under certain conditions there are also methods for obtaining solutions to a congruence modulo powers of a prime given the solution of the congruence modulo the prime.

Finally, if we know solutions to the congruences modulo all the prime power divisors of m, then from the solution to this simultaneous congruence system we can get a solution to the congruence modulo m.

We have solved the congruences of degree 1 in all generality (Theorem 4). The solution of second-degree congruences is already a much more difficult problem, and historically, their complete solution required a long time and the effort of a number of distinguished mathematicians. During the eighteenth century several mathematicians observed, based on extensive calculations, essentially equivalent relationships to the so-called reciprocity theorem, which we will formulate and prove below. It was GAUSS who observed it independently and first succeeded in proving it as well.

In the following, we consider only congruences of degree 2, and among these, only those with prime modulus.

The general second degree congruence can be reduced to the simpler congruence

$$x^2 \equiv c \pmod{p}, \tag{24}$$

essentially the problem of extracting a square root modulo p. It is immediately clear that the condition for the existence of the square root for real numbers, that the number should be nonnegative, is useless here, for the simple reason that every residue class has positive representatives and negative ones as well.

The case $c = -1$ ($\equiv p - 1$ (mod p)), for example modulo 5 and 29 has solutions $(\pm 2)_5$ and $(\pm 12)_{29}$, respectively, and Theorem 15 guarantees a solution modulo every prime of the form $4k + 1$. On the other hand, the statement of Exercise 23 shows that there are never solutions modulo a prime of the form $4k + 3$. The question of the existence of a solution to congruence (24) was one of the central problems during the early development of number theory.

The trivial cases of $c \equiv 0, x \equiv 0$ (mod p) we exclude in the remainder of this section, as well as the trivial case where $p = 2$.

If congruence (23) has a solution, then we say that c is a *quadratic residue* modulo p. If there is no solution, then it is called a *quadratic nonresidue*. Whether or not a number c (its residue class) is a quadratic residue is called c's *quadratic character*.

If we choose for x the residues of least absolute value, we get that the congruence has a solution if c is congruent to one of

$$(\pm 1)^2, (\pm 2)^2, \ldots, (\pm (p-1)/2)^2,$$

and otherwise there is no solution. It is easy to see that these numbers are pairwise incongruent modulo p. We leave the proof of this to the reader. This observation shows that exactly half of the residue classes relatively prime to p are quadratic residues, and the other half are not. With the help of this simple observation it is not difficult to show that multiplication of quadratic characters is similar to multiplication of ± 1. We formulate this in the following theorem:

Theorem 20 *The product of two numbers is a quadratic residue if both of the factors are quadratic residues or if both are nonresidues; the product is a quadratic nonresidue if one of the factors is a quadratic residue and the other a nonresidue.*

To make notation simpler, we will denote $(p - 1)/2$ by p_1. We need to show three things. The first is obvious. If a and b are quadratic residues, i.e., there are numbers x and y such that

$$x^2 \equiv a \pmod{p}, \qquad y^2 \equiv b \pmod{p},$$

then

$$(xy)^2 \equiv ab \pmod{p},$$

and ab is also a quadratic residue.

Now let k_0 be an arbitrary quadratic residue ($k_0 \not\equiv 0$ (mod p)) and let

$$k_1, k_2, \ldots, k_{p_1}, n_1, n_2, \ldots, n_{p_1} \tag{25}$$

be a reduced residue system, where the k's are quadratic residues, the n's nonresidues. Multiplying the system by k_0 we get a new reduced residue system, since k_0 is relatively prime to p. In this system

$$k_0 k_1, k_0 k_2, \ldots, k_0 k_{p_1}, k_0 n_1, k_0 n_2, \ldots, \ k_0 n_{p_1}$$

we know that the first p_1 elements are quadratic residues. The remaining elements are therefore nonresidues. Thus we have proved that if a number is a quadratic residue, then multiplying it by any nonresidue yields a nonresidue. This proves the third claim of the theorem.

To prove the second claim, let n_0 be an arbitrary quadratic nonresidue, and multiply the residue system (25) by n_0. We get the system

$$n_0 k_1, n_0 k_2, \ldots, n_0 k_{p_1}, n_0 n_1, n_0 n_2, \ldots, \ n_0 n_{p_1},$$

where the first p_1 elements are nonresidues, as we have just shown. It thus follows that the second p_1 elements are all quadratic residues, showing that the product of two quadratic nonresidues is a quadratic residue. This completes the proof of the theorem.

41. Theorem 20 suggests the introduction of a function, written $\left(\frac{c}{p}\right)$ and called the *Legendre symbol*. Its value is 1 if c is a quadratic residue modulo p, -1 if c is a quadratic nonresidue, and 0 if c is divisible by p.[9] The statement from Theorem 21 can be written, using the Legendre symbol, in the form

$$\left(\frac{ab}{p}\right) = \left(\frac{a}{p}\right)\left(\frac{b}{p}\right).$$

This relation clearly also holds if either a, b, or both are divisible by p. From the definition, it also follows that if c is not divisible by p, then $\left(\frac{c^2}{p}\right) = 1$.

42. EULER found a formula for the quadratic character of a number, which we express here with the help of the Legendre symbol:

Theorem 21 (Euler's Lemma) *If p is an odd prime, then for every c,*

$$\left(\frac{c}{p}\right) \equiv c^{(p-1)/2} \pmod{p}.$$

The case $c \equiv 0 \pmod{p}$ can be disregarded. In all other cases, the theorem says that c is either a quadratic residue or a nonresidue depending on whether the $(p-1)/2$ power of c is congruent to 1 or -1, respectively, modulo p.

We remark that Fermat's theorem follows immediately from Euler's lemma by squaring both sides.

We present a proof of this theorem without using either Wilson's theorem or Fermat's theorem, and in this way we also give a new proof of each of these theorems. Let c be an arbitrary number not divisible by p. Denote by $x_i, i = 1, 2, \ldots, p-1$, the smallest positive number such that

$$i x_i \equiv c \pmod{p}. \tag{26}$$

---

[9] The function and its representation are in no way related to the fraction c/p.

From Theorem 4 it follows that the solutions to these congruences are distinct residue classes. We also see that if $x_i = j$, which means that $ij \equiv c \pmod{p}$, then $x_j = i$.

If c is a quadratic nonresidue, so that there is no y satisfying

$$y^2 \equiv c \pmod{p}, \tag{27}$$

then every congruence in (26) appears twice, with the order of the factors on the left interchanged. If we take only one of each of these and multiply them together, we get $(p-1)!$ on the left-hand side. In this way, if c is a quadratic nonresidue, then

$$(p-1)! \equiv c^{(p-1)/2} \pmod{p}.$$

If y is a solution to (27), then $p - y$ is also a solution, and there are no others. Choose y between 0 and p. Then $p - y$ is also between 0 and p. If we disregard these two cases, then every remaining pair in (26) occurs twice. The two values of i omitted are y and $p - y$, which satisfy

$$y(p - y) \equiv -c \pmod{p}.$$

From each congruence appearing twice, we take one and multiply them together, and then multiply their product by this last congruence, getting again $(p-1)!$ on the left, so if c is a quadratic residue, then

$$(p-1)! \equiv -c^{(p-1)/2} \pmod{p}.$$

The question now remains as to what the left-hand side is congruent to. We know that 1 is a quadratic residue for all p, and substituting in for c we get

$$(p-1)! \equiv -1 \pmod{p},$$

which is Wilson's theorem. Substituting this in for the left-hand side of the previous two congruences, we complete the proof of Euler's lemma.

43. With Euler's lemma we now have a method for determining whether or not a number is a quadratic residue for a prime modulus. Since we need only the residue modulo p of the powers, this method is useful if the numbers concerned are not too large but still not very easy to compute. Let us determine whether or not 30 is a quadratic residue modulo 103. By Euler's lemma, we need to compute 30^{51} modulo 103. All of the following congruences are modulo 103:

$30^2 = 900 \equiv -27,$ $30^{32} \equiv 576 \equiv -42,$

$30^4 \equiv (-27)^2 = 729 \equiv 8,$ $30^{48} = 30^{16+32} \equiv -24(-42) = 1008 \equiv -22,$

$30^8 \equiv 64 \equiv -39,$ $30^{50} \equiv -27(-22) \equiv 594 \equiv -24,$

$30^{16} \equiv 1521 \equiv -24,$ $30^{51} \equiv -720 \equiv 1.$

The congruence $x^2 \equiv 30 \pmod{103}$, therefore, has a solution. None of the results we have so far help us in searching for the solutions (in our case $(37)_{103}$ and $(66)_{103}$), and such a simple method as the Euclidean algorithm in the case of first-degree congruences cannot be given.

It is interesting that if we know a number n that is a quadratic nonresidue modulo p, then there is a fairly good algorithm to find the solutions. The smallest such n is not very likely larger than a power of $\log p$ and can thus be computed in a realistic amount of time by a computer.[10] It has only been proved, however, that for every positive number h and sufficiently large p there exists such an n that is no larger than $p^{1/(4\sqrt{e})+h}$, and if this is the best possible upper bound, it does not follow that there is a good method to find the solution using a computer. If we know a primitive root, then we can make an index table, and with this it is an easy task to solve the congruence. The best known upper bound for the least primitive root[11] is $p^{1/4+h}$. This bound cannot be significantly improved, and this method, therefore, is also not very useful for finding a solution.

44. In one case Euler's lemma easily provides a general result. This is the case of $c = -1$. If the exponent is even, i.e., p is of the form $4k + 1$, then the power is 1; if the exponent is odd, i.e., p is of the form $4k + 3$, then the power is -1. This gives us the general result that -1 is a quadratic residue for all primes of the form $4k + 1$ and a quadratic nonresidue for primes of the form $4k + 3$.

We remark that Theorem 15 and Exercise 23 provide this result in a different way. From Euler's lemma we cannot hope to get such nice general results for other values of c. As the numbers get larger, it becomes increasingly difficult to calculate the Legendre symbol from this lemma.

Recalling the first proof of the Euler–Fermat theorem, we can try to simplify the task of determining whether or not a number c, not divisible by p, is a quadratic residue. Consider the residues with smallest absolute value of the following numbers:

$$c, 2c, \ldots, ((p-1)/2)c. \tag{27}$$

Then only the numbers $1, 2, \ldots, (p-1)/2$, appear, of course possibly with negative signs. This leads to the following congruences:

$$ic \equiv e_i b_i \pmod{p} \quad (i = 1, 2, \ldots, (p-1)/2), \tag{28}$$

where e_i is either 1 or -1 and $1 \leq b_i \leq (p-1)/2$.

We will show that no two of the b's are equal. It is obvious that no two can be the same and have the same sign, since we are multiplying half of a reduced residue system by a number relatively prime to p. We need to show, therefore, that there are no indices i and j for which

---

[10] Remark of I. Z. Ruzsa, personal communication.

[11] D. Burgess: *Mathematika* 4 (1957), pp. 106–112.

$$e_i b_i + e_j b_j = 0.$$

This means that

$$ic + jc \equiv 0 \pmod{p}, \quad \text{and hence} \quad p \mid (i+j)c.$$

This is impossible, since we assumed c not to be divisible by p, and i and j are such that $1 \le i + j \le (p-1)$. It then follows that

$$b_1 b_2 \cdots b_{(p-1)/2} = ((p-1)/2)!.$$

After the corresponding sides of the congruences (28) have been multiplied together, the product above occurs on both sides. Let m be the number of congruences in (28) with a negative sign. Then we have

$$((p-1)/2)! c^{(p-1)/2} \equiv (-1)^m ((p-1)/2)! \pmod{p}.$$

None of the factors are divisible by p, and we can simplify. With the help of Euler's lemma, we can write our results as

$$\left(\frac{c}{p}\right) \equiv (-1)^m \pmod{p}.$$

Since the difference between the two sides can be 2, 0, or -2, and the modulus is at least 3, the two sides must be equal. This important result is known as Gauss's lemma:

Theorem 22 (Gauss's Lemma) *Let p be an odd prime and c an integer not divisible by p. Consider the number of residues of smallest absolute value of the numbers $c, 2c, \ldots, ((p-1)/2)c$ that are negative. Then c is a residue or nonresidue according to whether this number is even or odd, respectively.*

45. With the help of this theorem, for any number we can determine for which primes it is a quadratic residue and for which primes it is a nonresidue.

In the case of 2, for example, we need to determine how many of the numbers $2, 4, \ldots, p-1$, have a negative number as their residue of smallest absolute value modulo p. In other words, we only need to count the number of even integers between $p/2$ and p, or expressed in yet a different way, the number of integers in the interval $(p/4, p/2)$.

It is useful to write p in the form $8k + r$, where r is 1, 3, 5, or 7. We then need to determine whether the number of integers in the interval $(2k + r/4, 4k + r/2)$ is even or odd. The endpoints are not integers, and we may omit the subinterval $(2k + r/4, 4k + r/4)$ of length $2k$, since there are $2k$, i.e., an even number, of integers in this interval, and it suffices to investigate the interval $(4k + r/4, 4k + r/2)$.

We also note that there is the same number of integers in the interval $(r/4, r/2)$, the previous interval translated by $4k$. For each of the four possible values of r, the number of integers in this interval is 0, 1, 1, and 2, respectively. Thus 2 is a quadratic residue in the first and the last cases, and a nonresidue in the two middle cases. We may formulate this as follows: *the number 2 is a quadratic residue for primes of the form $8k \pm 1$, and a nonresidue for primes of the form $8k \pm 3$.* Notice that this depends only on the residue class of p modulo 8.

46. As a second example let us determine for which primes p the number 5 is a quadratic residue. To do this, we need to determine which of the numbers $5, 10, \ldots, 5(p-1)/2$, have a negative number as their residue of smallest absolute value modulo p. All of these numbers are smaller than $5p/2$, and we need to count the number of integers divisible by 5 in the intervals

$$(p/2, p) \quad \text{and} \quad (3p/2, 2p).$$

This is the same as counting the number of integers in the intervals

$$(p/10, p/5) \quad \text{and} \quad (3p/10, 2p/5).$$

The case $p = 5$, of course, is omitted, and therefore the endpoints are not integers.

It is helpful to write p in the form $20k + r$. Since p is prime, r can now take on the values 1, 3, 7, 9, 11, 13, 17, or 19. For these numbers we need to determine whether the number of integers in each of the intervals

$$(2k + r/10, 4k + r/5) \quad \text{and} \quad (6k + 3r/10, 8k + 2r/5)$$

is odd or even. The subintervals

$$(2k + r/10, 4k + r/10) \quad \text{and} \quad (6k + 3r/10, 8k + 3r/10)$$

may be omitted, since there is an even number of integers in these, and it is enough to consider the remaining parts

$$(4k + r/10, 4k + r/5) \quad \text{and} \quad (8k + 3r/10, 8k + 2r/5),$$

or by a translation, even the intervals

$$(r/10, r/5) \quad \text{and} \quad (3r/10, 2r/5).$$

The numbers of integers in the intervals for the above choices of r are 0, 1, 1, 2, 2, 3, 3, and 4, respectively. The parity of these numbers depends only on the value of r and is independent of k. We can formulate our results in the following statement: The number 5 is a quadratic residue for primes p that upon division by 20 have a remainder of ± 1 or ± 9; for those primes with remainder ± 3 or ± 7, it is a nonresidue. More simply, we formulate this as follows: The number 5 is a quadratic residue for primes of the form $10k \pm 1$ and a nonresidue for primes of the form $10k \pm 3$.

47. As a third example, we examine when -3 is a quadratic residue, i.e., we determine for which primes p there is an even number of numbers among $-3, -6, \ldots, -3(p-1)/2$ whose remainder of smallest absolute value upon division by p is negative. The prime $p = 3$ is, of course, excluded. We need to count the number of multiplies of 3 between $-3p/2$ and $-p$ and also between $-p/2$ and 0, where 0 does not count.

Reflecting the numbers through 0, we have the intervals $(0, p/2)$ and $(p, 3p/2)$, which contain the same number of multiples of 3. The number of these is $[p/6] + [p/2] - [p/3]$. Write p in the form $12k + r$, where r can be 1, 5, 7, or 11, for p a prime. The values of the sum for the four cases are

$$2k + 6k - 4k = 4k; \qquad\qquad 2k + 6k + 2 - 4k - 1 = 4k + 1;$$
$$2k + 1 + 6k + 3 - 4k - 2 = 4k + 2; \qquad 2k + 1 + 6k + 5 - 4k - 3 = 4k + 3.$$

We conclude that -3 is a quadratic residue for primes of the form $12k + 1$ and $12k + 7$, or more simply for primes of the form $6k + 1$; for all others, i.e., primes of the form $6k - 1$, it is a nonresidue. We will use this result in the following sections, and hence we have highlighted it here.

Lemma *The number -3 is a quadratic residue for primes of the form $6k+1$ and a nonresidue for primes of the form $6p - 1$.*

48. It is clear that for any c we may decide in a similar way for which prime moduli the number c is a quadratic residue. Investigating different values of c, we saw that whether or not it is a quadratic residue modulo a prime p depends only on the residue of p modulo $4c$. This is not by chance. We will restrict ourselves to the case where c is positive (although the case for c negative is very similar).

We need to determine how many multiples of c fall in the intervals $((2t-1)p/2, tp)$, where t is a positive integer not larger than $c/2$ (the largest occurring multiple is $c(p-1)/2 < cp/2$).

The endpoints cannot be multiples of c, since we consider only those primes that are not divisors of c, and the starting points are not even integers because $t < c$.

The question is then how many integers lie in the intervals

$$\left(\frac{(2t-1)p}{2c}, \frac{tp}{c} \right).$$

It is suitable to write p in the form $4ck + r$, where k is an integer and $1 \le r \le 4c$, and we are interested in how many integers lie in the intervals

$$\left(2(2t-1)k + \frac{(2t-1)r}{2c}, 4tk + \frac{tr}{c} \right).$$

We again disregard the intervals

$$\left(2(2t-1)k + \frac{(2t-1)r}{2c}, 4tk + \frac{(2t-1)r}{c} \right),$$

since they each contain an even number of integers. The remaining intervals

$$\left(4tk + \frac{(2t-1)r}{2c}, 4tk + \frac{tr}{c} \right)$$

contain as many integers as the intervals

$$\left(\frac{(2t-1)r}{2c}, \frac{tr}{c} \right),$$

and this depends only on r, not on k, and this is what we wanted to show.

49. The multiplicative property of the Legendre symbol gives us a way to write every Legendre symbol as a product of terms of the form $\left(\frac{p}{q}\right)$, where q is a positive odd prime and p is a positive prime or -1. We have already settled the cases $\left(\frac{2}{q}\right)$ and $\left(\frac{-1}{q}\right)$, so it remains only to determine $\left(\left(\frac{p}{q}\right)\right)$ for arbitrary odd primes p and q. The search for such a general method was a long one, as we have mentioned.

EULER determined the quadratic residues for many different prime moduli, and he noticed two things: first, that the quadratic character of a positive number c modulo a prime p depends only on the residue of p modulo $4c$; second, that primes with residues r and $4c - r$ (equivalently $-r$) give the same result in determining the quadratic character of c, since we saw in the cases of 2 and 5. (In the second observation, it is essential for c to be positive, as we saw in the case of $c = -3$ that this statement is false.)

The proof of the first of these facts followed from Gauss's lemma. EULER, of course, did not have this at his disposal, and without it he did not succeed in proving his conjectures. We will show how the second observation also follows from Gauss's lemma.

In defining the quadratic character, LEGENDRE was the first person to state the so-called *reciprocity theorem*, equivalent statements of which had been conjectured by others. If p and q are distinct (positive) primes, then the Legendre symbols $\left(\frac{p}{q}\right)$ and $\left(\frac{q}{p}\right)$ have the same value, except in the case that both p and q are of the form $4k + 3$, in which case one symbol is 1 and the other -1.

GAUSS independently conjectured the theorem and at the age of 19 gave a rather complicated proof using induction. Throughout his life he often returned to this theorem and gave a total of seven proofs, all using different methods. We will see that the reciprocity theorem follows from EULER's two observations, and we will now prove the second one.

50. In the proof of the first of EULER's observations we already saw that to determine whether a number c is a quadratic residue modulo a prime of the form $4ck + r$ depends only on whether the number of integers in intervals of type $((2t - 1)r/2c, tr/c)$ is even or odd. For primes of the form $4ck - r$, we need to investigate the corresponding intervals for $r' = 4c - r$.

If the two quadratic characters agree, then that means that the sum of the integers in the first and second cases is even. We will show that this is true even for the pair of intervals of the two systems corresponding to the same value of t. Writing r' in place of r, we get the intervals

$$\left(4t - 2 - \frac{(2t - 1)r}{2c}, 4t - \frac{tr}{c} \right).$$

Reflecting the interval through the point $2t$ and taking the mirror image, we get the interval

$$\left(\frac{tr}{c}, 2 + \frac{(2t - 1)r}{2c} \right),$$

which contains the same number of integers as before because this reflection sends integers to integers.

This interval together with the original is an interval of length 2, and the endpoints are not integers; hence the interval contains 2 integers, and we have proved EULER's second observation:

Theorem 23 *For a positive number c, whether or not it is a quadratic residue modulo a prime depends only on the residue of the prime modulo $4c$; furthermore, those primes with residues r and $4c - r$ agree, or more simply, those with residues r and $-r$ agree.*

51. We can now prove the reciprocity theorem using Theorem 23 and the known properties of the Legendre symbol. Let p and q be two odd primes. Let us first investigate the case where one of the primes is of the form $4k + 1$ and the other $4k + 3$. We may write $p + q = 4c$, where c is a positive integer, and we have

$$\left(\frac{p}{q} \right) = \left(\frac{4c - q}{q} \right) = \left(\frac{4c}{q} \right) = \left(\frac{4}{q} \right) \left(\frac{c}{q} \right) = \left(\frac{c}{q} \right),$$

because the Legendre symbol is periodic with respect to the lower number, is multiplicative, and has value 1 when the upper number is a square.

In the same way we can show that $\left(\frac{q}{p} \right) = \left(\frac{c}{p} \right)$. By hypothesis, the residues of p and q modulo $4c$ are opposites, and by the second part of Theorem 23 we may conclude that

$$\left(\frac{p}{q} \right) = \left(\frac{c}{q} \right) = \left(\frac{c}{p} \right) = \left(\frac{q}{p} \right).$$

If p and q have the same remainder modulo 4, then call the larger of the two p and write $p - q = 4c$ for a positive integer c. We may then write

$$\left(\frac{p}{q}\right) = \left(\frac{4c+q}{q}\right) = \left(\frac{4c}{q}\right) = \left(\frac{4}{q}\right)\left(\frac{c}{q}\right) = \left(\frac{c}{q}\right)$$

and

$$\left(\frac{q}{p}\right) = \left(\frac{p-4c}{p}\right) = \left(\frac{-4c}{p}\right) = \left(\frac{-1}{p}\right)\left(\frac{4}{p}\right)\left(\frac{c}{p}\right) = \left(\frac{-1}{p}\right)\left(\frac{c}{p}\right).$$

By the second part of Theorem 24 we conclude that $\left(\frac{c}{p}\right) = \left(\frac{c}{q}\right)$ and that $\left(\frac{p}{q}\right) = \left(\frac{-1}{p}\right)\left(\frac{q}{p}\right)$. If p and q have residue 1 modulo 4, then we conclude that $\left(\frac{p}{q}\right) = \left(\frac{q}{p}\right)$. Otherwise, if they have residue 3, we conclude that $\left(\frac{p}{q}\right) = -\left(\frac{q}{p}\right)$. We can express our results in the following formula:

Theorem 24 (Reciprocity Theorem) *If p and q are distinct odd primes, then*

$$\left(\frac{p}{q}\right)\left(\frac{q}{p}\right) = (-1)^{\frac{1}{2}(p-1)\cdot\frac{1}{2}(q-1)}.$$

The exponent here is odd only if both primes are of the form $4k+3$. In this case one of the Legendre symbols is 1 and the other is -1. In all other cases the two Legendre symbols are either both 1 or both -1.

The previous results for the cases of -1 and 2 are useful in applications, and we recall them here:

Supplementary Theorem *For p a positive odd prime,*

$$\left(\frac{-1}{p}\right) = (-1)^{(p-1)/2}, \qquad \left(\frac{2}{p}\right) = (-1)^{(p^2-1)/8}.$$

For the case of 2, this formula expresses our results. The exponent is

$$\frac{1}{2}\frac{(p-1)}{2}\frac{(p+1)}{2},$$

and the numbers in parentheses are consecutive even integers; hence only one is divisible by 4. The exponent is therefore even if this factor is also divisible by 8, i.e., when p is of the form $8k \pm 1$, as we stated the result earlier.

As an application of our results, let us decide whether or not the congruence

$$x^2 \equiv 611 \pmod{1009}$$

has solutions; in other words, let us determine $\left(\frac{611}{1009}\right)$ (here $1009 = 16 \cdot 63 + 1$ is a prime and $611 = 13 \cdot 47$):

$$\left(\frac{611}{1009}\right) = \left(\frac{13}{1009}\right)\left(\frac{47}{1009}\right) = \left(\frac{1009}{13}\right)\left(\frac{1009}{47}\right) = \left(\frac{-5}{13}\right)\left(\frac{22}{47}\right)$$

$$= \left(\frac{-1}{13}\right)\left(\frac{5}{13}\right)\left(\frac{2}{47}\right)\left(\frac{11}{47}\right) = +1 \cdot \left(\frac{13}{5}\right)(+1)\left(-\left(\frac{47}{11}\right)\right)$$

$$= -\left(\frac{3}{5}\right)\left(\frac{3}{11}\right) = -\left(\frac{5}{3}\right)\left(-\left(\frac{11}{3}\right)\right) = \left(\frac{2}{3}\right)\left(\frac{2}{3}\right) = 1.$$

The congruence is therefore solvable (the solutions of it are $(653)_{1009}$ and $(356)_{1009}$).

Exercises:

32. Which of the following congruences are solvable and which are not:

$$x^2 \equiv 533 \pmod{1607}, \qquad x^2 \equiv 1238 \pmod{2011},$$

$$x^2 \equiv 3772 \pmod{5183}, \qquad 12x^2 \equiv 23 \pmod{113}.$$

$(5183 = 71 \cdot 73$, the other moduli are prime.)

33. Prove that the quadratic character of a negative number c modulo a prime p depends only on the residue of p modulo $4c$, as we proved it does for positive numbers c.

34. Determine the quartic (fourth power) residues for the moduli 11, 13, 17, and 19. Prove that modulo primes of the form $4k + 3$, quadratic residues are also quartic residues; modulo primes of the form $4k + 1$, only half of the quadratic residues are quartic residues.

35. Prove that if a is an integer and u and v are positive integers, then

$$(a^u - 1, a^v - 1) = |a^{(u,v)} - 1|.$$

36. Prove that if p is a prime and $(k, p - 1) = d$, then the congruence

$$x^k \equiv 1 \pmod{p}$$

has d solutions.

37. Let p be a prime and c a number not divisible by p.
   (a) Prove that if the congruence

$$x^k \equiv c \pmod{p} \tag{29}$$

   is solvable, then

$$c^{(p-1)/(k,p-1)} \equiv 1 \pmod{p}, \tag{30}$$

   and $(k, p - 1)$ residue classes satisfy congruence (29).
   (b) Prove that if the condition (30) is satisfied, then the congruence (29) has a solution.

38. Prove that for all primes p of the form $4k - 1$,

$$\left(\frac{p-1}{2}\right)! \equiv (-1)^m \pmod{p},$$

where m is the number of quadratic residues between $p/2$ and p.

3. Rational and Irrational Numbers. Approximation of Numbers by Rational Numbers (Diophantine Approximation)

1. As we have already mentioned, the numbers we know the best are the integers. These occur very infrequently among the real numbers. Quotients of these, the rational numbers, occur much more often, but as we will see, in a certain sense, they make up only an insignificant part of the real numbers. It is rather surprising, however, that it can be quite difficult to decide whether or not a given number is rational.

We know how to do computations with rational numbers, especially with integers, or with (finite) decimals. It is not surprising that we should try to approximate real numbers by rational numbers, and if possible, by those with small denominators. The investigations concerning such questions are called *Diophantine approximation.*

In this chapter we start by determining whether certain numbers, or certain types of numbers, are rational. Among these, we are also going to distinguish two types. We will then investigate how real numbers may be approximated by rational numbers with good accuracy, where the accuracy is measured in terms of the size of the denominator.

2. The different number systems give very practical ways of representing numbers. (See Exercises 7, 9, 10, in Chapter 1.) In Fibonacci's book, which appeared in 1201, he introduced Arabic numerals and their usefulness in performing operations. Following Fibonacci, many excellent mathematicians endeavored to popularize their use, but in many instances the use of Roman numerals was enforced. In fact, the acceptance or Arabic numerals into common usage in society was very slow. In earlier times, numbers were represented. by letters; for a long time, a large part of Europe used Roman numerals, and for certain purposes they are still used today.

Arabic numerals had appeared by the beginning of the thirteenth century and with them the decimal number system. It was, however, only in the fifteenth and sixteenth centuries that these started to come into common use, and it took centuries still for them to become fully accepted.

Earlier, we mentioned finite decimals. It is quite easy to see (Exercise 1) that finite decimals arise from fractions whose denominator, when the fraction is written in reduced form, is not divisible by any primes other than 2 and 5.

We can obtain the decimal notation of a rational number by dividing the numerator by the denominator, using long division. Once we have used the digits of the numerator, we put a decimal point to the right of the quotient, and then adding a zero to the right of the remainder we continue the division. The above-mentioned claim states that this process will never result in a remainder of 0 if the divisor is divisible by a prime other than 2 and 5.

If the decimal is not finite, then the remainder at each step is always positive and smaller than the divisor. There are only finitely many numbers of this type (one fewer than the divisor), and therefore sooner or later (at the latest after writing as many zeros as the size of the divisor), a remainder must occur that has already occurred. Previously, we also added a 0 to this and continued the division; thus we see that the same digit occurs in the quotient as earlier, as well as the same remainder. So it is clear that the quotients and remainders repeat periodically.

Rational numbers, therefore, have a decimal notation that is either finite or has a string of digits that infinitely repeats itself; the latter ones are called *repeating decimals*.

3. What do we mean, however, by an infinite decimal? We can think of it as adding infinitely many terms and never reaching the end. The idea of a summation really loses meaning in the general case of infinitely many summands, and we know that we cannot even associate a real number to every infinite sum. Many of these infinite sums, which we will call by the name *infinite series*, can be summed, and for these we say that the series *converges*. Infinite decimals are series that converge, and if the decimal is repeating, then we can easily determine the sum (see Fact 4). We need to be careful in dealing with convergent series, since not all the properties of normal addition hold.[1] Those relations that we need we will use without proof, as we did already in the first chapter.

The finite decimals are just a certain class of fractions (those whose denominator is a power of 10). Is it possible to find a fraction for every periodic decimal so that the infinite decimal arises from the fraction?

To start with, we will not consider those infinite decimals that from some point on are all 9's. It is easy to see that these do not arise from long division (see Exercise 2). For the sake of simplicity, we will work only with what are called *pure periodic* decimals, those whose periodic interval starts right after the decimal point. Essentially the same method will work for those numbers that have a nonrepeating part preceding the periodic part.

We will investigate the following pure periodic decimal:

$$u = \overline{a_1 a_2 \ldots a_n . b_1 b_2 \ldots b_k \, b_1 b_2 \ldots b_k \, b_1 \ldots},$$

---

[1] See, for example, W. W. SAWYER: *What Is Calculus About?*, S.M.S.G.; Math. Assoc. of America, 1961.

where the a_i and b_i are digits, the dot after a_n is the decimal point, and the overline is there to indicate that we should consider the sequence of digits and not their product. We let a be the number before the decimal (whose digits are the a_i), and let b be the k-digit number (sequence) that repeats. By grouping the digits after the decimal point in groups of k, we may write u as

$$u = a + \frac{b}{10^k} + \frac{b}{10^{2k}} + \cdots.$$

Multiplying by 10^k changes only the integer part:

$$10^k u = 10^k a + b + \frac{b}{10^k} + \frac{b}{10^{2k}} + \cdots.$$

We can subtract u from this number, and since the decimal parts are the same, they cancel each other out, and the difference is

$$\left(10^k - 1\right) u = \left(10^k - 1\right) a + b.$$

So we have

$$u = \frac{(10^k - 1)a + b}{10^k - 1} = a + \frac{b}{10^k - 1}.$$

With this we see that u is a rational number.

4. We can, however, reach our goal without worrying about infinite series by showing that the decimal notation of the fraction $\left((10^k - 1) a + b\right) / \left(10^k - 1\right)$ is the infinite decimal we started with.

It is obvious that dividing $\left(10^k - 1\right) a$ by $\left(10^k - 1\right)$ we get $a = \overline{a_1 a_2 \ldots a_n}$, and that when we subtract off the term corresponding to a_n, we are left with $b = \overline{b_1 b_2 \ldots b_k}$ as the remainder. This is smaller than the divisor, since we specifically excluded the case of infinitely repeating 9's.[2] Let us now append a 0 to the remainder:

$$\overline{b_1 b_2 \ldots b_k 0} = 10^k b_1 + \overline{b_2 \ldots b_k 0} = \left(10^k - 1\right) b_1 + \overline{b_2 \ldots b_k b_1}.$$

Thus the quotient is b_1 and the remainder is the last term. This remainder can be obtained from the the previous one by removing the first digit and writing it at the end. If we continue this for a total of k times, then k times we remove the first digit and write it at the end; after this we will return to the original remainder. The quotient is always the first digit of the remainder; hence its digits are $b_2, b_3, \ldots, b_k$, and the same decimal arises that we started with. We formulate our results here:

Theorem 1 *Every rational number can be written as either a finite decimal or an infinite periodic decimal with infinitely many digits different from 9.*

---

[2] If this were the case, the last digit of the quotient would be $a_n + 1$ (carrying the 1, if necessary), and in this case u would be an integer.

Conversely, every finite decimal and every infinite periodic decimal with infinitely many digits different from 9 is the decimal representation of a rational number.

We also include the integers here as finite decimals, where they are understood to have a decimal part of length 0.

Exercises:

1. Prove that those fractions whose denominators are not divisible by any primes other than 2 and 5 have a finite decimal notation.
2. Prove that an infinite decimal whose digits are all 9's after a certain point cannot occur from long division of a fraction.
3. Prove that Theorem 1 holds for all periodic decimals.

5. It is easy to write aperiodic decimals; for instance

$$0.10100100010000\ldots,$$

where between each pair of consecutive 1's we write one more 0 than between the previous pair (in other words, we write down the consecutive powers of 10 after the decimal point). We may also write down the number whose decimal consists of all positive integers written in order:

$$0.123456789101112\ldots.$$

In both of these numbers there will occur arbitrarily many consecutive zeros, but after these, nonzero digits will also occur. If these decimals were recurring, then they would have periodic intervals that would fall in an interval of all zeros, hence the digits would all be zero from some point on, but this is not the case. Therefore, the two numbers are irrational.

6. The next theorem will give a method for creating infinitely many irrational numbers of the type above. We denote by $\langle a \rangle$ the sequence of the base-10 digits of the positive number a. (We assume the first to be different from 0.)

Theorem 2 *If the infinite sequence of distinct positive integers*

$$a_1, a_2, \ldots$$

leads to a rational number whose decimal notation is

$$0.\langle a_1 \rangle \langle a_2 \rangle \ldots, \tag{1}$$

then there is a number K such that for every positive n,

$$\sum_{i=1}^{n} \frac{1}{a_i} \leq K.$$

The first example above shows that the condition is necessary, but not sufficient, for such a number to be rational. We could conclude from the theorem that the second example above is irrational, since it is well known that the sequence of the sums

$$H_n = 1 + \frac{1}{2} + \frac{1}{3} + \cdots + \frac{1}{n}, \qquad n = 1, 2, \ldots,$$

grows beyond all bounds (Fact 6).

Turning to the proof of the theorem, the condition in the theorem says that the decimal (1) starts somewhere to be periodic. Let r be an index r for which all of $\langle a_r \rangle$ falls in the periodic part, and let the length of a period (the number of digits) be s. Then for $i \geq r$, the place of the first digit of a_i and the number of digits in a_i are determined; if the number of digits is t, then

$$a_i \geq 10^{t-1},$$

and the number of t-digit numbers among the a's occurring in the periodic part is at most s. Thus if $n \geq r$ and the largest a_i occurring up to a_n has u digits, then

$$\sum_{i=1}^{n} \frac{1}{a_i} = \sum_{i=1}^{r-1} \frac{1}{a_i} + \sum_{i=r}^{n} \frac{1}{a_i} \leq \sum_{i=1}^{r-1} \frac{1}{a_i} + \sum_{t=1}^{u} \frac{s}{10^{t-1}}$$

$$= \sum_{i=1}^{r-1} \frac{1}{a_i} + s \frac{1 - 1/10^u}{1 - 1/10} = \sum_{i=1}^{r-1} \frac{1}{a_i} + s \frac{10 - 1/10^{u-1}}{9}$$

$$< \sum_{i=1}^{r-1} \frac{1}{a_i} + \frac{10s}{9}.$$

The last sum can be chosen for K, since it does not depend on n, only on the number (1).

We will see (Theorem 5.3) that the sum of the reciprocals of the prime numbers diverges, thus as an application of this theorem we have the result that the number

$$0.2357111317\ldots,$$

of the form (1) where a_i is the ith prime number, is irrational.[3]

Exercise:

4. Let $b = (b_1, b_2, \ldots, b_s)$ be a sequence of numbers such that $0 \leq b_i \leq 9$. We say that a number a does not contain b if in base-10 notation, the elements of b do not occur in their given order as consecutive digits of a.

---

[3] For the theorem and its proof, see N. HEGYVÁRI: *The American Mathematical Monthly* **100** (1993), pp. 779–780, and L. E. TAYLOR: *ibid.* **101** (1994), p. 174.

(a) Prove that if $a_1, a_2, \ldots$ is a sequence of distinct numbers for which none of the elements contain b, then the sum

$$\sum_{i=1}^{n} \frac{1}{a_i}$$

is smaller than some constant not depending on n.

(b) Prove Theorem 2 using the statement in (a) (Hegyvári).

7. We used the fundamental theorem of arithmetic to prove that $\sqrt[k]{c}$ is irrational if c is an integer but not the kth power of an integer (Theorem 1.11). We will give proofs for some special cases of this result without referring to the fundamental theorem. Most of these proofs will be geometric.

The existence of irrational numbers was essentially known to the ancient Greeks, even though they did not call them by this name. They showed their existence geometrically by showing that certain distances were not commensurable with each other. Two intervals are called *commensurable* if they can both be measured off without remainder by repeatedly using one interval, called a *common scale*; otherwise, they are called *incommensurable*. Two distances are therefore commensurable if the ratio of their lengths is rational, and they are incommensurable if this ratio is irrational.

It is, of course, not possible to determine whether two intervals are commensurable just by looking at them, but some type of logical argument can be persuasive. One of the classic results of incommensurability shows that the side and diagonal of a square are incommensurable, which amounts to showing that $\sqrt{2}$ is irrational. We present this result here in a little more generality.

Theorem 3 *For every positive integer m, the number $\sqrt{m^2 + 1}$ is irrational.*

We prove this by showing that if ABC is a right triangle with leg AC having length m times that of leg AB, then AB and the hypotenuse BC are incommensurable.

Let us assume that there is a common scale of the two intervals. Draw the circle with center C and radius AC; let D be the point of intersection of the circle and the side BC (see Figure 1). We have now measured off the length $AC = CD = m\,AB$ onto the segment BC, and the common scale can be measured off evenly into this too. The common scale can therefore also be measured off into BD without remainder. The remaining part BD is smaller than AB, because

$$BC = BD + CD,$$
$$BC < AB + AC = AB + CD.$$

Draw the tangent to the circle at D, and let E be the point of intersection with the segment AB. Then

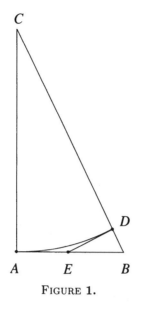

FIGURE 1.

$$AE = DE$$

since they are both tangents, and also the new right triangle DBE is similar to the original, since B is their common vertex; we see that

$$DE = m\,BD.$$

In this way, we have already measured off the length BD m times onto AB. We can measure the length $ED = m\,BD$ onto BE by drawing a circle with center E and passing through D. The remainder will now be smaller than BD. It is clear that this process will continue indefinitely.

With this observation we have a contradiction, since we assumed that there is an interval that can be measured off evenly into the two lengths, and we showed that it can be measured off evenly into smaller and smaller lengths, still without remainder. We thus produced an infinite sequence of positive integers that is strictly decreasing, which is impossible. The assumed common scale therefore cannot exist, and we have completed the proof of the theorem.

In the case $m = 1$, this is the classical proof that $\sqrt{2}$ is irrational.

8. The incommensurability of the two intervals could have been seen right from the first pair of similar triangles. We need only to realize that if two intervals are both multiples of a common scale then among all common scales, there is a largest one. If a common scale of two intervals is measured off r times into the first interval, then a larger common scale could be measured off only $r - 1$ times, $r - 2$ times, ..., twice, or once, without remainder. Thus the common scale is just an rth of the interval. We need only check the interval itself, half of it, a third of it, ... , an $(r-1)$st of it, and an rth of it, to determine which is the first that is a common scale. By our hypothesis, the last is a common scale, and we have thus shown the existence of a largest common scale.

Supposing now that AB and BC are commensurable, consider the largest common scale. This would also be a common scale of AB and CD, and furthermore of CD and DE. Enlarge the triangle DBE together with this common scale, so the triangle is the size of ABC. In this way we have a new common scale for AB and BC that is larger than the previous common scale, which was largest possible. This is a contradiction.

Exercises:

    5.   Give a geometric proof of the fact that for all positive integers m greater than 1, the number $\sqrt{m^2-1}$ is irrational.

    6.   Prove that the side of a regular decagon (10-sided polygon) is incommensurable with the radius of its circumcircle (in other words two unequal sides of an isosceles triangle whose central angle is $36°$ are incommensurable).

9. The fact that $\sqrt{2}$ is irrational can also be proved in an indirect way with an easy calculation. Let us assume that it is rational, i.e., it can be written as ℓ/k, where k and ℓ are integers. Then k and ℓ would satisfy the equation

$$2k^2 = \ell^2.$$

We may further assume that ℓ/k cannot be simplified. The left-hand side of the equation is even, and it follows that ℓ must also be even, for the square of an odd number is odd. We can write $\ell = 2m$ for some integer m. Substituting this into the equation and simplifying, we have $k^2 = 2m^2$. By the same argument as before, we see that k must be even, but then k and ℓ are both divisible by 2, contradicting the fact that they are relatively prime. Thus $\sqrt{2}$ cannot be rational.

10. Both types of proofs can be generalized. We will prove the following theorem first in a geometric way.

Theorem 4 *If c is a positive integer and not the square of an integer, then $\sqrt{c}$ is irrational.*

We will use the following known theorem: *The area of a square whose side length is that of the altitude of a right triangle is the same as the area of the rectangle whose side lengths are those of the two parts of the hypotenuse determined by the altitude*[4] (Figure 2).

---

[4] J. SURÁNYI: "Schon die alten Griechen haben es gewußt" in *Große Augenblicke aus der Geschichte der Mathematik* (RÓBERT FREUD, ed.) Akadémia Kiadó (1990), pp. 9–50; especially pp. 23–26 (in German).

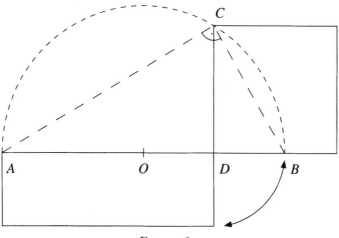

FIGURE 2.

We can then visualize $\sqrt{c}$ in the following way. Measure intervals $AD = c$ and $DB = 1$ on a line (D between A and B); construct a half circle whose diameter is AB and the perpendicular to line AB through the point D. Call the point of intersection of this line with the circle C (Figure 3). Then triangle ABC is a right triangle with right angle C, and we have

$$CD^2 = AD \cdot DB = c, \quad \text{i.e.,} \quad CD = \sqrt{c}.$$

The claim of the theorem is then that the two legs of the triangle BCD are incommensurable.

Measure off intervals of length 1 (the length of BD) onto CD starting from D. Then

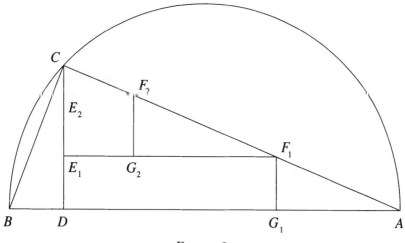

FIGURE 3.

$$DE_1 = E_1 E_2 = \cdots = E_{k-1} E_k = 1, \quad \text{and} \quad 0 < E_k C < 1 = BD,$$

because c is not the square of an integer. (In the figure, $k = 2$.) Construct lines perpendicular to CD through the points E_i, intersecting AC in points F_i, respectively. From F_1 construct a line perpendicular to AD, intersecting it in point G_1, and lines from the F_i perpendicular to $E_{i-1} F_{i-1}$ intersecting it in point G_i ($i = 2, 3, \ldots, k$). Then the triangles $F_1 A G_1$ and $F_{i+1} F_i G_{i+1}$ ($i = 1, 2, \ldots, k - 1$) are congruent to triangle BCD, and triangle $CF_k E_k$ is similar to BCD and smaller, since CE_k is shorter than BD.

We now assume that the theorem is false, i.e., that BD and CD have a common scale. Consider the largest common scale. This can be measured into AD and DE_k without remainder, since these are integer multiples of DB, and by our assumption, it can also be measured into $E_k C$. It could also be measured into $E_k F_k$ without remainder, since this is what remains from AD after measuring it into the parallel intervals $AG_1, F_1 G_2, F_2 G_3, \ldots, F_{k-1} G_k$ of length CD.

Enlarge the triangle $CF_k E_k$ with its largest common scale to the size of triangle BCD. With this we get a larger common scale for the two legs than the one we started with (which we assumed to be the largest). This is a contradiction. The two lengths are therefore incommensurable.

11. The geometric proof hints at an alternative arithmetic proof for the theorem. Following the construction above and writing down the ratios of the legs of the similar triangles, we get

$$\frac{F_k E_k}{E_k C} = \frac{AD - k\,DC}{DC - k\,DB} = \frac{c - k\sqrt{c}}{\sqrt{c} - k}, \qquad \frac{DC}{DB} = \sqrt{c}.$$

These ratios are equal, so we have

$$\sqrt{c} = \frac{c - k\sqrt{c}}{\sqrt{c} - k}.$$

This relationship holds for all c and k (independently of the previous geometric arguments), as long as the denominator is not zero.

Let us write $\sqrt{c}$ as u/v, where u and v are positive real numbers. Simplifying the fraction on the right side, we get

$$\frac{cv - ku}{u - kv}. \tag{2}$$

Let us now assume that $\sqrt{c}$ is rational and u/v is written with the smallest possible positive integers. Choose the integer k so that the denominator in (2) is positive and smaller than v:

$$0 \le u - kv \le v, \quad \text{and therefore} \quad 0 \le \frac{u}{v} - k = \sqrt{c} - k < 1.$$

Since we assumed that $\sqrt{c}$ is not an integer, we must choose $[\sqrt{c}]$ for k (see Fact 8). For such a choice of k we get a new fraction for $\sqrt{c}$ with a smaller denominator. This is a contradiction, and we have a new proof of the theorem.[5]

12. The proof of the following theorem runs along very different lines.

Theorem 5 *If m is an integer larger than 4, then $\tan(\pi/m)$ is irrational.*

The following proof was given by P. TURÁN (1910–1976) as a student. We will use the fact that $\tan m\alpha$ can be represented as the quotient of two polynomials in $\tan\alpha$ with integral coefficients. We write t instead of $\tan\alpha$ for simplicity. Then

$$\tan m\alpha = \frac{\binom{m}{1}t - \binom{m}{3}t^3 + \binom{m}{5}t^5 - \cdots}{1 - \binom{m}{2}t^2 + \binom{m}{4}t^4 - \cdots}$$

(of course, only finitely many of the coefficients in both the numerator and denominator are nonzero). The identity can be proved using induction.[6] In our proof we will use the above relationship only for odd m, and all we need from it is that for odd m we can write

$$\tan m\alpha = \frac{\pm t^m + P_m(t)}{1 + Q_m(t)},$$

where $P_m(t)$ and $Q_m(t)$ are both polynomials in t with integer coefficients and degrees less than m (possibly the 0 polynomial), and the constant term of $Q_m(t)$ is 0.[7] This we will prove by induction.

In the case $m = 1$, we have $\tan\alpha = t$, and $P_1(t) = Q_1(t) = 0$. Let us assume that the claim is true for some odd value k of m. Then

$$\tan(k+2)\alpha = \frac{\tan k\alpha + \tan 2\alpha}{1 - \tan k\alpha \tan 2\alpha} = \frac{\left(\pm t^k + P_k(t)\right)/(1 + Q_k(t)) + 2t/\left(1 - t^2\right)}{1 - 2t\left(\pm t^k + P_k(t)\right)/\left(1 - t^2\right)(1 + Q_k(t))}.$$

Multiplying out the denominators and simplifying, we can rewrite the fraction as

---

[5] Here the geometric proof led us to an arithmetic proof, which is, however, independent of the geometry. This is all the more the case because the arithmetic proof, due to R. DEDEKIND, arose much earlier. See *Stetigkeit und irrationale Zahlen*, Friedrich Vieweg & Sohn, Braunschweig (1872), pp. 12–13 (in German). The geometric proof is due to GY. HAJÓS (1912–1972) and was communicated verbally to the authors.

[6] This identity can also be proved by applying DE MOIVRE's theorem, which states that $\cos m\alpha + i\sin m\alpha = (\cos\alpha + i\sin\alpha)^m$, expanding this using the binomial theorem, and then solving for $\sin m\alpha / \cos m\alpha$ in terms of $\sin\alpha / \cos\alpha$.

[7] It is important for the proof of Theorem 5 that the coefficient of the highest-degree term in the numerator be ± 1. This is not true for m even, and that is why we treat the case of odd m separately.

$$\frac{\mp t^{k+2} \pm t^k + \left(1 - t^2\right) P_k(t) + 2t(1 + Q_k(t))}{1 + Q_k(t) - t\left(t(1 + Q_k(t)) + 2\left(\pm t^k + P_k(t)\right)\right)}.$$

We see by the inductive hypothesis that the parts of denominator and the numerator after the first term are of degree less than $k+2$; in the denominator the constant term is 1, since the constant term of $Q_k(t)$ is 0 and the last term is divisible by t. With this we have proved the claim.

With the help of this we prove Theorem 5 for odd m. For $\alpha = \pi/m$, we have that $\tan m\alpha = \tan \pi = 0$, so $t = \tan \pi/m$ is a root of the polynomial in the numerator of the expression above for $\tan m\alpha$. Write this polynomial in the form

$$P_m(t) = p_{m-1}t^{m-1} + p_{m-2}t^{m-2} + \cdots + p_1 t + p_0,$$

where the coefficients are integers and the degree is at most $m - 1$, as we just showed.

Assume that the theorem is false and $\tan \pi/m = u/v$ is a rational number. Further, assume that the fraction cannot be simplified further; hence u and v are relatively prime integers. Of course, $v \neq 0$. Then

$$\pm t^m + P_m(t) = \pm \left(\frac{u}{v}\right)^m + p_{m-1}\left(\frac{u}{v}\right)^{m-1} + p_{m-2}\left(\frac{u}{v}\right)^{m-2} + \cdots + p_0 = 0.$$

Multiplying both sides by v^m and factoring out a v from all terms containing it, we have

$$\pm u^m + v\left(p_{m-1}u^{m-1} + p_{m-2}u^{m-2}v + \cdots + p_0 v^{m-1}\right) = 0.$$

The number in parentheses is an integer, so we see that v is a divisor of the first term. Since we assumed u and v to be relatively prime, this is possible only if $v = \pm 1$, meaning that $t = \tan \pi/m$ is an integer.[8] This immediately leads to a contradiction for $m \geq 5$, since we know from the geometry that $0 < \tan \pi/m < 1$. This proves Theorem 5 for odd m. (We can include the case $m = 3$, since $\tan \pi/3 = \sqrt{3}$, and this we have already shown to be irrational.)[9]

13. We now consider the case where m is even. Let $m = 2^k n$, where n is odd and k is at least 1; actually, if $n = 1$, then we assume k to be at least 3 to satisfy the condition that $m > 4$.

If $\tan(\pi/m)$ were rational, then $\tan(\pi/(m/2)) = \tan(2\pi/m)$ would be rational because $\tan 2\alpha = 2 \tan \alpha/(1 - \tan^2 \alpha)$. If $n > 1$, then repeating this k times, we see that $\tan(\pi/n)$ would be rational, which we just proved is not true.

---

[8] In the last two paragraphs we have essentially shown that in an algebraic equation with integer coefficients, if the term of highest degree has coefficient 1, then among all rational numbers, only integers can be roots.

[9] Alternatively, we can show using geometry that $1 < \tan(\pi/3) < 2$. Using the above argument, we can show that if $\tan(\pi/3)$ is rational, then it must be an integer, and this is impossible.

If $n = 1$, then repeating the argument $k - 3$ times, it would follow that $\tan(\pi/8)$ is rational, but

$$\tan(\pi/8) = \sqrt{2} - 1$$

is irrational. This completes the proof for even m.

We have already seen that $\tan(\pi/3)$ is irrational. On the other hand, for $m = 1$ and $m = 4$, $\tan \pi = 0$ and $\tan(\pi/4) = 1$ are rational, and for $m = 2$, $\tan(\pi/2)$ has no meaning.

14*. In Fact 5 we referred to natural logarithms and their base e. This number can be defined as the sum of the series

$$1 + \frac{1}{1!} + \frac{1}{2!} + \frac{1}{3!} + \cdots + \frac{1}{n!} + \cdots .$$

To the first few decimal places, its value is 2.7182818. In what follows, we accept without proof that this series converges, and that the arithmetic operations with it used in the course of the proof are allowed; we also accept the formula for the sum of geometric series given in Fact 4. Using these, we will prove the following result.

Theorem 6 *The number e is irrational.*

If e were rational, then its denominator would divide $n!$ for any n at least as large as the denominator. Let us choose such an n. Then $n!\,e$ would be an integer. Multiply the above series termwise by $n!$. Multiplying the first $n + 1$ terms, we get an integer A_n; the rest of the terms can then be simplified by $n!$. Subtracting A_n we get

$$n!\,e - A_n = \frac{1}{n+1} + \frac{1}{(n+1)(n+2)} + \frac{1}{(n+1)(n+2)(n+3)} + \cdots .$$

By replacing each factor of every divisor by $n + 1$, we increase the sum. In this way we get an infinite geometric series with ratio $1/(n+1)$ whose sum is

$$\frac{1/(n+1)}{1 - 1/(n+1)} = \frac{1}{n}.$$

Hence we have

$$0 < n!\,e - A_n < \frac{1}{n} < 1.$$

This, however, is not true for any integer, and we conclude that $n!\,e$ cannot be an integer. This proves Theorem 6.

15. More than this can be said about e. There is no algebraic equation (polynomial) with integer coefficients, of however large degree, that has e as a root. The fact that e is irrational means only that e is not the root of an equation of the form $ax - b = 0$, where a and b are integers, $a \neq 0$. The irrational numbers of the form $\sqrt[k]{m}$ are roots of the algebraic equation $x^k - m = 0$, where k and m are integers, as was discussed in Theorem 1.11. Those numbers that are roots of algebraic equations with integer coefficients are called *algebraic numbers*; those that do not fall into this category are called *transcendental numbers*. The number e is transcendental, but the proof of this fact far exceeds the scope of this book. LIOUVILLE was the first to prove that there are transcendental numbers. The proof of this will be Exercises 12 and 13.[10]

Transcendental numbers do exist, and the proof mentioned in footnote 10 suggests that most numbers are transcendental. On the other hand, deciding whether a given number is transcendental is a hard problem; in fact, deciding whether it is rational can already be very difficult.

In addition to e, the ratio of the circumference of a circle to its diameter, π, has also been shown to be transcendental. This solved the more than 2000-year-old question of squaring the circle: It is not possible to construct, in the Euclidean sense, using a straightedge and compass, a square whose circumference, nor one whose area, equals that of a given circle.

It is still unknown whether the number $e + \pi$ is rational. A rather interesting question arises in connection with the infinite series

$$\zeta_k = 1 + \frac{1}{2^k} + \frac{1}{3^k} + \cdots + \frac{1}{n^k} + \cdots .$$

It can be shown that this series converges for any real number k greater than 1. For all even integers k, the sum can be given in closed form, and in fact this was done by EULER. The sum of the reciprocals of squares, for instance, is

$$\zeta_2 = 1 + \frac{1}{2^2} + \frac{1}{3^2} + \cdots + \frac{1}{n^2} + \cdots = \frac{\pi^2}{6}.$$

It easily follows from this that ζ_2 is transcendental, and similarly this can be deduced for ζ_{2j} for every positive integer j. On the other hand, it was only in 1979 that R. APÉRY proved that ζ_3 is irrational, a result that aroused worldwide attention;[11] others have since proved the same result using different techniques. The problem is still open for higher odd values of k.

It is also unknown whether the so-called *Euler–Mascheroni constant* is rational. This number appears in the investigation of several mathematical questions. It can be defined in the following way: The sum $H_n = 1 + \frac{1}{2} + \frac{1}{2} + \cdots + \frac{1}{n}$ is greater than $\log(n + 1)$ (see Fact 6), but not by much. The difference $H_n - \log(n + 1)$ is bounded for every n (for instance, it is smaller

---

[10] For a totally different (set-theoretic) proof of this fact, see G. H. HARDY, E. M. WRIGHT: *An Introduction to the Theory of Numbers*, Oxford (1960), Section 11.6.

[11] See A. VAN DER POORTEN: *Math. Intelligencer* **1**, 1979, pp. 195–203; F. BEUKERS: *Bull. London Math. Soc.* **11**, No. 3 (1979), pp. 268–272; E. REYSSAT: *Seminaire Delaunay-Pisot-Poitou 20ème Année*, 1978/79, *Théorie des Nombres*, Fasc. 1 No. 6.

than 1), and as n gets larger, the differences converge to a certain number. This number is the above-mentioned constant.

Exercises:

7. Prove that if $g \geq 2$ is an integer, then the two series

$$\sum_{n=0}^{\infty} \frac{1}{g^{n^2}} \quad \text{and} \quad \sum_{n=0}^{\infty} \frac{1}{g^{n!}}$$

both converge to irrational numbers.

8. Prove that numbers of the form

$$\frac{a_1}{1!} + \frac{a_2}{2!} + \frac{a_3}{3!} + \cdots ,$$

where $0 \leq a_i \leq i - 1$ $(i = 2, 3, \dots)$, are rational if and only if starting from some i on all the a_i's are either equal to 0 (in which case the sum is finite) or all are equal to $i - 1$.

9*. Using the fact that

$$\frac{1}{e} = 1 - \frac{1}{1!} + \frac{1}{2!} - \frac{1}{3!} + \cdots = \sum_{i=0}^{\infty} \frac{(-1)^i}{i!},$$

prove that e is not the root of any second-degree polynomial with integer coefficients.

10*. Let the infinite sequences $a_1, a_2, \dots$ and $b_1, b_2, \dots$ be such that each of their elements is either a 2 or a 3. Define

$$A_n = \frac{1}{a_1} + \frac{1}{a_1 a_2} + \cdots + \frac{1}{a_1 a_2 \cdots a_n},$$
$$B_n = \frac{1}{b_1} + \frac{1}{b_1 b_2} + \cdots + \frac{1}{b_1 b_2 \cdots b_n}.$$

Prove that if a_i and b_i are different for some i, then for all $n \geq i$, the quantity $|A_n - B_n|$ is greater than a constant independent of n. (In other words, a number can be represented in at most one way by an infinite series

$$\frac{1}{a_1} + \frac{1}{a_1 a_2} + \frac{1}{a_1 a_2 a_3} + \cdots .) \tag{3}$$

11*. (Continuation) Prove that a number of the form (3) is rational if and only if the sequence of the a_i's is periodic from some point on.

12*. Prove that if α is the root of an nth-degree polynomial with integer coefficients that cannot be written as the product of integer polynomials of smaller degree,[12] then there exists a number k such that, with the

---

[12] Polynomials of this type are called *irreducible*.

exception of at most finitely many fractions u/v, the following relation holds:

$$\left| \alpha - \frac{u}{v} \right| > \frac{k}{v^n},$$

where u and v are integers (Liouville).

13. (Continuation) Prove that the following infinite series converges to a transcendental number:

$$\frac{1}{2^{1!}} + \frac{1}{2^{2!}} + \cdots + \frac{1}{2^{n!}} + \cdots$$

(Liouville).

16. So far we have dealt with irrational and transcendental numbers. Still, we know the rational numbers much better.

In the previous section we mentioned that we do not know, for instance, whether the Euler–Mascheroni constant or the numbers ζ_{2k+1} for $k \geq 2$ are rational. Whichever the case may be, calculations with these numbers are not easy based on their (rather complicated) definition. In the introduction we mentioned that in cases like this it would be convenient to approximate real numbers by rational numbers with small denominators.

It is known that the decimal representation of the Euler–Mascheroni constant is $0.5772156649\ldots$. If we approximate this to only two decimal places by $0.58 = 58/100 = 29/50$, then the deviation of this fraction from the constant is less than 0.003. However, the rational number $11/19 = 0.57894\ldots$, written with much smaller integers, differs by less than 0.002 from the constant. If we consider fractions with denominators slightly larger than 100, we see that $71/123 = 0.57723577\ldots$ already differs by less then 0.000021 from the constant.

In general, if one is required to approximate a real number α by a fraction with a given denominator v (for example, with denominators 100 or $1\,000\,000$), then we consider the two multiples of $1/v$ that surround α, and choose the one that is close to α. This cannot be at a distance greater than half of the interval of length $1/v$ away from α, so this procedure yields a fraction u/v for which

$$\left| \alpha - \frac{u}{v} \right| \leq \frac{1}{2v}.$$

We cannot improve this for an arbitrary v, since α may fall right in the middle of the interval $1/v$, or arbitrarily close to the middle.

The above examples show that among the fractions whose denominators are not greater than v, we can find ones that are much closer to α than $1/(2v)$. Examining the product of this difference with v, we would like to know how small we can make the number

$$v \left| \alpha - \frac{u}{v} \right| = |v\alpha - u|$$

if we consider all integers u and v, with v positive and not greater than a given value q.

We can already give a bound much smaller than $\frac{1}{2}$ to this question.

Theorem 7 *For any real number α and any positive integer q, it is possible to find integers u and v such that $v \neq 0$ and*

$$|v\alpha - u| \leq \frac{1}{q+1} \quad and \quad 1 \leq v \leq q.$$

Applying this theorem to the approximation by the fraction u/v gives

$$\left|\alpha - \frac{u}{v}\right| = \frac{1}{v}|v\alpha - u| \leq \frac{1}{v(q+1)} < \frac{1}{v^2}.$$

In this light it is not so surprising that in the above example we could find such close approximations.

The only restriction on u in Theorem 7 is that u must be an integer. That is, the theorem states something about the distances between the numbers $\alpha, 2\alpha, \ldots, q\alpha$ and the integers closest to them; it is not important which integer this is. It is easier to understand these quantities if we consider them as arcs measured on the perimeter of a circle of circumference 1 starting from a point P_0, measure the arcs $\alpha, 2\alpha, \ldots, q\alpha$ onto the perimeter, not worrying if we wind around the circle more than once. This amounts to disregarding the integer part of a number.

Call the positive direction that in which we measured these arcs onto the perimeter of the circle and let the negative direction be its opposite (see Figure 4, where $q = 12$).

Denote the points on the perimeter in the order they were measured by $P_1, P_2, \ldots, P_q$. Together with P_0, we have divided the perimeter into $q+1$ arcs, and the length of the shortest such arc is not greater than $1/(q+1)$. Let $\overparen{P_r P_s}$ be a shortest arc. This is the length of the arc, independent of the direction discussed above. (We do not exclude the case that the points P_r and P_s coincide, in which case α is a rational number that can be written as a fraction with denominator not exceeding q.) We now assume that α is irrational and $r > s$, say. We can then formulate our observation

$$\overparen{P_r P_s} = |\overrightarrow{P_0 P_r} - \overrightarrow{P_0 P_s}| \leq \frac{1}{q+1},$$

where $\overrightarrow{P_0 P_r}$ and $\overrightarrow{P_0 P_s}$ are the lengths of the arcs between P_0 and P_r (respectively P_s), measured in the positive direction. These values are the fractional parts of $r\alpha$ and $s\alpha$, respectively, which we can write as the number minus its integer part. In other words

$$|r\alpha - [r\alpha] - s\alpha + [s\alpha]| = |(r-s)\alpha - ([r\alpha] - [s\alpha])| \leq \frac{1}{q+1}.$$

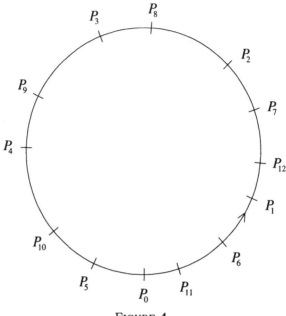

FIGURE 4.

Here $0 \leq s < r \leq q$, i.e., $0 < r - s \leq q$, and defining $u = [r\alpha] - [s\alpha]$ and $v = r - s$ we have integers u and v satisfying the claim of Theorem 7.[13]

If α is rational, then with a large enough q, we get α itself as the approximating fraction, and there is no better approximation than this. This way, the series of approximating values in the theorem is finite. Theorem 7 would not be worth much if this situation were true for irrational α as well. Luckily, it is not hard to see that this is not the case.

Theorem 8 *Let α be an irrational number. There exist infinitely many fractions u_n/v_n $(n = 1, 2, 3, \dots)$ that approximate α to within $1/v_n^2$.*[14]

Applying Theorem 7 here with $q_1 = 1$ we get the first such fraction u_1/v_1 ($v_1 = 1$ and u_1 is the closest integer to α). We now choose an integer q_2 so that the corresponding fraction u_2/v_2 guaranteed by Theorem 7 is closer to α than u_1/v_1 was. For this it suffices to choose q_2 so that $1/(q_2+1) < |v_1\alpha - u_1|$ is satisfied.

Proceeding further in this fashion, if we already have found u_i/v_i, then we choose the next integer q_{i+1} satisfying $1/(q_{i+1} + 1) < |v_i\alpha - u_i|$. This is always possible, since α is irrational, and this way none of $v_i\alpha - u_i$ can be 0. The next fraction u_{i+1}/v_{i+1} corresponding to q_{i+1} in Theorem 7 satisfies

---

[13] The proof also shows that equality holds if and only if the P_i's divide the perimeter of the circle into equal arcs of length $1/(q+1)$, that is, if α is a rational number whose denominator in its simplest form is $q + 1$.

[14] For α rational, the expanded forms of the fraction α (e.g., for $\frac{1}{2}$ we have $\frac{1}{2}, \frac{2}{4}, \frac{3}{6}$, etc.) provide these infinite approximating values.

$$|v_{i+1}\alpha - u_{i+1}| < \frac{1}{q_{i+1} + 1} < |v_i\alpha - u_i|,$$

and hence this new fraction is different from any of the previous ones. As mentioned earlier in the discussion of Theorem 7, for these fractions u_n/v_n the following inequality holds:

$$\left|\alpha - \frac{u_n}{v_n}\right| < \frac{1}{v_n(q_n + 1)} < \frac{1}{v_n^2}.$$

17. We have shown that for every irrational number there are infinitely many fractions close to it in the sense above. This leads to further questions. How do we find such a fraction? Is it possible to give a better approximation for every irrational number in the sense of Theorem 7?

There are different ways to answer these questions. The technique used in the proof of Theorem 8 gives one possible method. We saw that in the case where we return to P_0 (when α is rational), then the points are equally spaced around the circle (the vertices of a regular polygon).

Even before the sequence possibly returns to P_0—in the case where α is irrational it never returns—the points are distributed fairly regularly. We saw in the proof of Theorem 7 that the distances between neighboring points first appear on an arc with the first point P_0. We can see that the distance between two neighboring points (measured in term of arc length) is equal to the distance from P_0 to the closest point on one of its two sides, or to the distance between the two points that are closest to P_0, each on a different side; hence it can take on only 3 possible values.

If for a given n we know the indices of the points neighboring P_0, then the order of the indices for all points can be determined. Based on these observations, we can answer the questions raised above. This method was worked out mainly by V. T. Sós, and she also gave important applications of it.[15] (See the exercises below.)

We will handle these questions in the next chapter based on a different geometric interpretation.

Exercises:

14. Using the same notation from the proof of Theorem 7, prove that if in measuring off n-times an arc of length α on a circle with unit circumference, the sequence does not land on P_0 again, and P_r and P_s are the closest neighbors of P_0 on each side, then if we continue measuring off arcs, P_{r+s} is the first point that will fall between P_r and P_s.

---

[15] See, for example, V. T. Sós: *Acta Math. Hung.* **VIII** (1957), pp. 462–472, and **IX** (1958), pp. 229–241, and *Annales Univ. Sci. Eötvös L. Sect. Math. 1* (1958), pp. 127–134.

15. (Continuation) Using the notation of the previous exercise, let P_v be the closest neighbor of P_u in the direction giving the shorter arc from P_0 to P_s. Show that only the following cases arise:

(a) If $u \geq r$, then $v = u - r$,

(b) If $u \leq n - s$, then $v = u + s$,

(c) If $n - s < u < r$, then $v = u - r + s$.

We note that all u between 1 and n fall into exactly one case. The statement of the problem implies the three-distance theorem we mentioned, since we see that $\widehat{P_u P_v}$ is equal to $\widehat{P_r P_0}$ in case (a), it is equal to $\widehat{P_0 P_s}$ in case (b), and it is equal to $\widehat{P_r P_s}$ in case (c).

There are always values of u that satisfy case (a) and that satisfy case (b), but if $n = r + s - 1$, then there is no u that satisfies case (c). In this case, there are only two different distances that appear between neighboring points.

18. The solution of Exercise 12 furnishes LIOUVILLE's theorem, which states that if α is the root of an irreducible polynomial of degree n with integer coefficients, then there exists a number k such that with the exception of at most finitely many fractions u/v, the following relation holds:

$$\left| \alpha - \frac{u}{v} \right| > \frac{k}{v^n}.$$

A. THUE improved this statement in 1909, showing that for any k, no matter how large, the conclusion still holds. From this it follows that if f is an irreducible polynomial of degree at least 3 with integer coefficients and c a nonzero integer, then the equation

$$F(x, y) = y^n f(x/y) = c$$

can have only finitely many integer solutions. Equations of this type are called *Thue equations*. THUE proved more than this. He showed that for any positive (small) ϵ, the inequality

$$\left| \alpha - \frac{u}{v} \right| > \frac{1}{v^{n/2 + \epsilon}}$$

has only finitely many solutions.

19. K. F. ROTH proved in 1955 that if α is an irrational algebraic number, then for arbitrary positive ϵ, the inequality

$$\left| \alpha - \frac{u}{v} \right| > \frac{1}{v^{2 + \epsilon}}$$

can have at most finitely many fractions as solutions.[16] For this result he received a Fields Medal[17] at the 1958 International Congress of Mathematicians in Edinburgh.

With the new methods of this deep result, new research directions were undertaken. Still many problems remain unsolved. Is it possible to replace v^ϵ in the denominator by some other function of v (for example $\log v$) that goes to infinity more slowly than the exponential? ROTH's techniques apparently cannot be applied in this direction, whereas this would be interesting because of its applications.

Further questions remain, such as whether there exists an algebraic number α such that for all positive ϵ there are integers u and v satisfying the inequality

$$\left| \alpha - \frac{u}{v} \right| < \frac{\epsilon}{v^2}.$$

Nothing is yet known in this direction.

There is a corollary of ROTH's results, similar to that for THUE's which we mentioned, concerning the number of integer solutions of algebraic equations in two variables. Namely, if $f(x)$ is an irreducible polynomial of degree n with integer coefficients, $F(x,y) = y^n f(x/y)$, and $G(x,y)$ is a polynomial of degree at most $n-3$ with integer coefficients, then the equation $F(x,y) = G(x,y)$ can have at most finitely many integer solutions (x,y).

These results are quite significant, since in the theory of Diophantine equations there had been very few general results before these, and most of the equations required new and original ideas.

20. Many important generalizations of THUE's and ROTH's theorems were given later. We will mention two important ones. C. L. SIEGEL (1926) showed that if $m \geq 2$ is an integer, and $f(x)$ is a polynomial of degree at least 3, with integer coefficients, and having all distinct roots, then the so-called *superelliptic* equation

$$y^m = f(x)$$

has only finitely many solutions with integers x and y.

In order to state the second consequence, let $p_1, \ldots, p_s$ be distinct primes. Denote by S the set of rational numbers that in reduced form have a denominator that is not divisible by primes other than $p_1, \ldots, p_s$. We call such rational numbers *S-integers*, and those for which the numerator is also not divisible by primes different from those given are called *S-units*. We then call an equation of the form

---

[16] See *Mathematica*, **2** (1955), pp. 1–20, and p. 168. The reader will find references in the article to the earlier results.

[17] Fields Medals are given to outstanding mathematicians for work they do by the age of 40. They are awarded every four years at the International Congress of Mathematicians, usually to two to four mathematicians; they are considered one of the most prestigious awards in mathematics.

$$au + bv = c$$

an *S-unit equation* whenever a, b, and c are integers different from 0, and we are interested in S-unit solutions. (We may assume that a, b, c are positive, relatively prime, and not divisible by any of the primes $p_1, \ldots, p_s$.) More precisely, we are looking for integers $x_1, \ldots, x_s, y_1, \ldots, y_s$ that are solutions to

$$a p_1^{x_1} \cdots p_s^{x_s} + b p_1^{y_1} \cdots p_s^{y_s} = c.$$

In 1933, K. MAHLER showed that even this has only finitely many solutions.[18]

The theorems mentioned above have few applications, since their proofs are not constructive; they do not give a method to find a bound above which there is no solution. In the 1960s, A. BAKER worked out a method by which it is possible to give bounds on the solutions of many Diophantine equations.[19] His work was recognized in 1970 at the International Congress of Mathematicians in Nice, France, where was awarded a Fields Medal. We will mention some of the many results that appeared in the wake of BAKER's work.

21. BAKER applied his methods to many problems, among them Thue equations and superelliptic equations. The latter were significantly generalized by B. BRINDZA. The best known bounds for the Thue equations were given by BUGEAUD and GYŐRY.[20] Their result states that every solution x, y satisfies

$$\max(x, y) < B^{cH^{2n-2}} (\log H)^{2n-1},$$

where

$$B = \max(|b|, 3), \qquad c = (n+2)^{18(n+1)},$$

and $H \, (> 3)$ is an upper bound for the absolute values of the coefficients of f.

Using BAKER's method, again BUGEAUD and GYŐRY obtained the best known bounds, that

$$\max(|x_1|, \ldots, |x_s|, |y_1|, \ldots, |y_s|) < (3(s+1))^{5(s+5)} P(\log P)^{s+1} \log H,$$

where P is the largest of the primes $p_1, \ldots, p_s$, and $H = \max(a, c, b, 3)$.

The bounds given above for n, H, s, and P can most likely be greatly improved, but this appears to be a very difficult problem. In principle, these bounds enable us to find all solutions, and also to prove that certain equations

---

[18] See for example L. J. MORDELL: *Diophantine Equations*, Academic Press, 1969. This work also gives a broad survey of the variety of methods used to solve different equations.

[19] For the method and its applications, see A. BAKER: *Transcendental Number Theory*, 3rd edition, Cambridge, 1990, and T. N. SHOREY, R. TIJDEMAN: *Exponential Diophantine Equations*, Cambridge, 1986.

[20] Y. BUGEAUD, K. GYŐRY: *Acta Arithmetica* **74** (1996), pp. 67–80 and pp. 273–292.

do not have any integer solutions, but in practice, these algorithms take too much time to execute, even on the fastest computers. For certain equations it has been possible to further reduce the bounds to a point where all solutions could be found by a computer. In the past 15 years, many people, including A. PETHŐ, B. DE WEGER, B. TZANAKIS, M. MIGNOTTE, I. GAÁL, M. POHST, have obtained results of this type.

The experience to date suggests that in the theory of Diophantine equations it is very rare that an equation has "large" solutions, or "many" solutions. This was supported by a result of E. BOMBIERI and W. M. SCHMIDT (1987), which states that the number of relatively prime solutions x, y to a Thue equation is at most cn^{t+1}, where c is a positive absolute constant, and t is the number of distinct prime divisors of b. It is interesting to note that this bound does not depend on the coefficients of f.

In 1984, J. H. EVERTSE showed that the number of solutions of an S-unit equation is at most $3 \cdot t^{2s+1}$, which is also independent of the coefficients a, b, and c. P. ERDŐS, C. L. STEWART, and R. TIJDEMAN showed that this is not far from the best possible bound, for if we let $a = b = c = 1$, $p_1, \ldots, p_s$ be the first s primes, and s be large enough, then the S-unit equation has more than

$$2^{(s/\log s)^{1/2}}$$

solutions. EVERTSE, GYŐRY, STEWART, and TIJDEMAN also showed that with only finitely many exceptions, there are at least two solutions to the above equation for triples a, b, c relatively prime to each other and also to $p_1, \ldots, p_s$, and hence the bound cannot in general be improved.

22. In the 1970s W. M. SCHMIDT extended ROTH's theorem to the simultaneous approximation of many algebraic numbers, even to linear forms with algebraic coefficients. He was able to use his methods for numerous equations in arbitrarily many unknowns and S-unit equations with arbitrarily many elements to show that there are only finitely many solutions, and in fact to obtain an upper bound on the number of solutions, and he was also able to give the structure of the entire solution set.[21] Unfortunately, SCHMIDT's methods do not give a bound on the size of the solutions.

The problem of determining the best upper bound for Diophantine approximation is the subject of the next chapter.

---

[21] For a good survey of recent investigations and results, see A. BAKER (ed.): *New Advances in Transcendence Theory*, Cambridge, 1988, chapter 10; W. M. SCHMIDT: *Diophantine Approximations and Diophantine Equations*, 1991; and K. GYŐRY: *Publ. Math. Debrecen* **42** (1993), pp. 65–101.

4. Geometric Methods in Number Theory

1. In the proofs of the previous chapters we often used geometrical considerations. We will present one more such proof. Wilson's theorem (Theorem 2.14) was easily proved in two different ways using congruences. We will now give a proof of it that does not use congruences.

The claim of the theorem, without reference to congruences, is that if p is a prime, then $(p-1)! + 1$ is divisible by p. For $p = 2$ this is clearly true, so we will consider only odd primes p. By Fact 1 we know that $(p-1)!$ is the number of permutations (reorderings) of the numbers $1, 2, \ldots, p-1$. We will represent every permutation as a directed closed polygon inscribed in a circle. Let $A_0, A_1, \ldots, A_{p-1}$ be the vertices of a regular inscribed p-gon. To a permutation $j_1 j_2 \ldots j_{p-1}$, we will associate the closed figure whose directed edges are from A_0 to A_{j_1}, from A_{j_1} to $A_{j_2}, \ldots$, from $A_{j_{p-2}}$ to $A_{j_{p-1}}$, and then from $A_{j_{p-1}}$ back to A_0. (Figure 1 corresponds to $p = 7$ and the permutation 265143.)

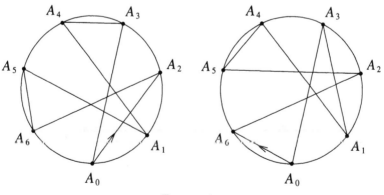

FIGURE 1.

If T_0 is such a closed figure, then by rotating T_0 repeatedly by $360/p$ degrees, we get a sequence

$$T_0, T_1, \ldots, T_n, \ldots$$

of closed figures. (The first rotation of the figure produces the permutation 625413; see Figure 1.) It is clear that $T_p = T_0$, and from then on the figures repeat themselves periodically; it is possible, however, that a closed figure occurs earlier. Let T_k be the first that is the same as a previous T_i. Then $i = 0$, for if it were larger, T_{k-1} would have already been the same as T_{i-1}.

The distinct closed figures $T_0, T_1, \ldots, T_{k-1}$ repeat periodically, and those for which the index is a multiple of k are the same as T_0. Thus p is divisible by k, and hence k is either 1 or p. With these rotations we have partitioned all the closed figures into sets of size 1 and sets of size p.

Those closed figures that are already mapped onto themselves with the first rotation can be obtained by drawing a segment from A_0 to some A_k and rotating. In this way we always get a closed figure with p vertices. If the figure were to become closed after fewer than p steps, then this closed figure and its isolated copies would contain the p copies of $A_0 A_k$. Thus the number of sides of one closed figure would divide p, and since 1 side is not possible, our claim holds true. The number of polygons that are the same with each rotation is therefore the number of ways of drawing the first edge from A_0, which is $p - 1$. The remaining

$$(p - 1)! - (p - 1) = (p - 1)! + 1 - p$$

closed figures are divided into groups of size p. This means that p divides $(p - 1)! + 1$, proving the theorem.

2. Further questions arise from this theorem. If the prime is 2, 3, or 5, then $(p - 1)! + 1$ is 2, 3, or 25, respectively. Can $(p - 1)! + 1$ be a power of p for other primes? In Chapter 7 we will show that this does not happen for other primes. It is possible, however, that $(p-1)!+1$ is divisible by p^2, for example $13^2 \mid 12! + 1$. It is still unknown whether there are infinitely many primes for which this holds.[1]

The geometric proof of Fermat's theorem and the proof above are similar not only in the sense that they both use closed figures inscribed in the circle.[2] In the exercises that follow we formulate the common thread of this argument and give examples of further applications.

Exercises:

1. Let S be an n-element set and **T** a one-to-one mapping of S onto itself. We write the repetition of **T** m times as $\mathbf{T}^m$.

---

[1] This property is shared by 563 and by no other primes up to 200 000. (Personal communication from I. Z. RUZSA.)

[2] In the first case we illustrated variations with repetitions and their cyclical permutations; in the second, we place a 0 in front of the permutations of the numbers $1, 2, \ldots, p - 1$; we reformulate the rotation as follows: To every element of a permutation we add 1, replacing p by 0; we then cyclically permute the numbers to put 0 at the beginning. With these reformulations, however, the proof would lose its suggestiveness.

(a) Prove that if c is an arbitrary element of S, then the sequence

$$c, \mathbf{T}c, \mathbf{T}^2c, \ldots, \mathbf{T}^mc, \ldots$$

is periodic, and if t is the shortest period and $\mathbf{T}^nc = c$, then $t \mid n$.

(b) Show that the two proofs mentioned above can be seen as applications of this theorem.

2*. Prove using the claim in the previous exercise that if p is a prime and $1 \le a \le p - 1$, then $\binom{p}{a}$ (the number of a-element subsets of a p-element set; see Fact 2) is divisible by p.

3*. Prove using the theorem in Exercise 1 that if the positive integers $a, b, c,$ and d satisfy $ab = cd$, then there exist positive integers $r, s, t,$ and u satisfying

$$a = rs, \quad b = tu, \quad c = rt, \quad d = su.$$

(This is the four number theorem, Theorem 1.2.)

4*. Use the theorem in Exercise 1 to prove the following version of Fermat's theorem: If p is a prime and $1 \le a \le p - 1$, then $p \mid a^{p-1} - 1$ (Theorem 2.11′).

We note that in proving the above exercises it was necessary to use only the fact that prime numbers are indecomposable, and in this way every result can be used to show that prime numbers have the prime property (Theorem 1.7).[3] This and the more general Euclid's lemma appeared in the first chapter as consequences of the four number theorem.

3. Until now, the geometric proofs we have presented were related to the circle, and at the end of the previous chapter we mentioned that these types of applications can be much further developed. Another branch of geometry that has many applications to number theory, especially to Diophantine approximation, is that of geometric lattices. This was first investigated by H. MINKOWSKI, and it is called the geometry of numbers.[4] In what follows we will be concerned with lattices.

Theorem 3.7 from the previous chapter admits, e.g., the following geometric interpretation. We associate to every pair of integers (u, v) a point in the plane whose coordinates are

$$x = v\alpha - u, \quad y = v. \tag{1}$$

---

[3] See K. HÄRTIG, J. SURÁNYI: *Periodica Math. Hung.* **6** (1975), pp. 235–240.

[4] In addition to his articles, he wrote two books that thoroughly deal with this subject: *Geometrie der Zahlen* (Leipzig, 1897) and *Diophantische Approximationen* (Leipzig, 1907). A modern treatment is given in J. W. S. CASSELS: *An Introduction to the Geometry of Numbers* (Springer, 1959), and further in the encyclopedic work P. GRUBER, C. G. LEKKERKERKER: *Geometry of numbers* (North Holland, 1987).

If $\alpha = 0$, then we get all points in the plane with integer coordinates. If $\alpha \neq 0$, then let us consider those points for which $v = 0$ and u is an arbitrary integer. These are all the points on the x-axis with integer coordinates, hence a sequence of points on a line at equally spaced intervals of length 1. For $u = 0$ we get the points that arise by repeatedly measuring off the vector $(\alpha, 1)$. These are the points on the line e through the origin with slope $1/\alpha$ occurring at equally spaced intervals. For another fixed u and an arbitrary integer v, the points in (1) arise from the points of e by translation by the vector $(-u, 0)$. If now v is fixed and u is arbitrary, these points arise by translating the points on the x-axis by the vector $(v\alpha, v)$.

Points of the form (1) are therefore the intersection of two families of parallel lines. In each family, the parallel lines occur at equally spaced intervals (this distance can be different in the two families). The two families together divide the plane into congruent parallelograms all having the same orientation. Two such parallelograms can intersect in either a side or a vertex (Figure 2).

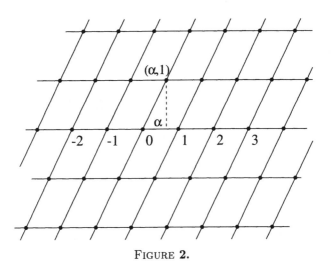

FIGURE 2.

The structure created by the lines is called a *parallelogram lattice*, the points of intersection are called *lattice points*, and all these points together are called a *point lattice*. The fact that one of the families (and in the case $\alpha = 0$, both families) consisted of lines parallel to a coordinate axis is only the consequence of the special problem. This will not be assumed in the sequel.

The parallelogram lattice uniquely determines the point lattice, but as we see in Figure 3, the converse is not true.

4. From our observations above, a parallelogram lattice is determined by two intersecting lines in the plane and two distances. We then consider the family of lines that are parallel to the first line and at a distance an integral

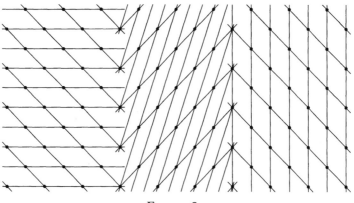

FIGURE 3.

multiple of the first distance; furthermore, we will consider as well the family of lines that are parallel to the second line and at a distance an integral multiple of the second distance. We call these lines the *lattice threads*. The intersection points of the lines determine the lattice points of the lattice. The parallelograms determined by neighboring pairs of lines are called *base parallelograms*.

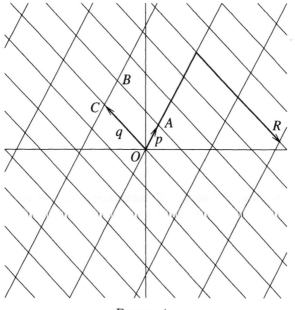

FIGURE 4.

Put a coordinate system on the plane in such a way that the origin O is a lattice point. Let $OABC$ be a base parallelogram, and let p and q be the vectors determined by the sides OA and OC, respectively (Figure 4). From O we can arrive at an arbitrary lattice point R by first traversing along line OA until we reach the line parallel to OC containing R, and then we continue along this line to R. The first shift is an integral multiple of $\boldsymbol{p}$ (possibly negative), the second an integral multiple of $\boldsymbol{q}$. In this way we can write the vector $\boldsymbol{r} = \overrightarrow{OR}$ as a sum of multiples of $\boldsymbol{p}$ and $\boldsymbol{q}$ in the following way:

$$\boldsymbol{r} = u \cdot \boldsymbol{p} + v \cdot \boldsymbol{q}. \tag{2}$$

In this way, u and v are uniquely determined, for if we obtain somehow a representation $\boldsymbol{r} = u'\boldsymbol{p} + v'\boldsymbol{q}$, then we have

$$(u - u')\boldsymbol{p} = (v - v')\boldsymbol{q},$$

and this is possible only if $u = u'$ and $v = v'$, since $\boldsymbol{p}$ and $\boldsymbol{q}$ are not parallel.

It is clear that for all integers u and v, moving by the vector (2) brings the origin onto a lattice point.

Let the coordinates of $\boldsymbol{p}, \boldsymbol{q}$, and $\boldsymbol{r}$ be (k, l), (m, n), and (x, y), respectively. From (2) we have

$$x = uk + vm, \qquad y = ul + vn. \tag{2'}$$

Here $d = |kn - lm|$ is the area of a base parallelogram, as the values $(0,0)$, $(1,0)$, and $(0,1)$ of (u, v) furnish the coordinates of the points O, A, and C, and the area of the triangle determined by these three vertices of a base parallelogram is half of d. This is obviously nonzero.

Theorem 1 *The lattice points of a parallelogram lattice are those points given by (2) for two nonparallel vectors $\boldsymbol{p}$ and $\boldsymbol{q}$ and two integers u and v, and they can also be expressed by (2') for real numbers k, l, m, and n. Here $d = |kn - lm|$ is the area of a base parallelogram.*

These lattices have many interesting properties. Two basic ones are mentioned here:

Basic Properties. I. *If we translate a lattice so that one lattice point is moved onto another, then the entire lattice is moved onto itself.*

II. *There is a positive number such that the distance between (distinct) lattice points is never smaller than this number.*

To verify property I, we need to see that every lattice point moves onto a lattice point, and every lattice point is the image of a lattice point. This is obvious once we realize that there is a line from each family of lattice threads that goes through the image of the translated point, and that these lines are parallel to the lines through the original point. The lines in these families are parallel and evenly spaced, and thus we see that the families are mapped onto themselves; hence the parallelogram lattice and the point lattice are also mapped onto themselves.

Property II is also easy to verify. For an arbitrary lattice point, we consider the four base parallelograms of which it is a vertex. These together determine a parallelogram twice the size of a base parallelogram with the chosen point in the middle and no other point inside. The smaller of the two heights of a base parallelogram is therefore a distance with the required property.

We can infer further properties of lattices using these basic properties, some of which are quite surprising.

5. The lines that pass through at least two lattice points we will call *lattice lines*. The lattice threads are examples of lattice lines. We similarly define *lattice vectors*, *lattice intervals*, and *lattice polygons* to be those whose vertices are lattice points. We can rephrase the first property as follows: Translation by a lattice vector maps the lattice onto itself.

If we consider two points of a lattice line e, then by the second property, there can be only finitely many lattice points on e between these two points. We can assume that A and B are neighboring lattice points on this line (Figure 5). Translating the lattice by $\overrightarrow{AB}$ maps e onto itself, and B onto a lattice point B'. Between B and B' there is no lattice point, since such a point would be the image of a lattice point between A and B, but we assumed that there is no such point. Similarly, B' is mapped onto a neighboring lattice point B'', etc. Translation in the opposite direction moves A onto a neighboring lattice point A', A' onto a neighboring lattice point A'', etc.

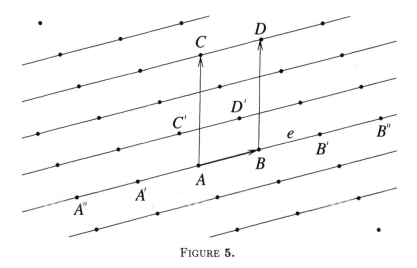

FIGURE **5.**

Consider a lattice point C not on e. By the two properties, translation by vector $\overrightarrow{AC}$ maps e onto a line parallel to it, which has lattice points at distance AB. Call D the image of B. The parallelogram $ABDC$ contains only finitely many lattice points. There are therefore only finitely many lattice lines between AB and CD parallel to e, since every such line has an interval of length AB falling in the parallelogram, and every such interval contains at least one lattice point.

If $C'D'$ is one of the neighboring lines parallel to e, then translation by the vector $\overrightarrow{AC}$ maps $C'D'$ into one of its neighboring lines, this line into one of its neighboring lines, and so on. In the opposite direction, e is mapped into its other neighboring lattice thread, which is at the same distance as e is from $C'D'$, and so on.

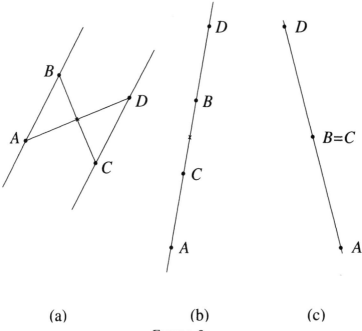

(a)   (b)   (c)

FIGURE 6.

We can formulate our observations as follows: If A, B, and C are lattice points and $\overrightarrow{AB} = \overrightarrow{CD}$, then D is also a lattice point, and $ABDC$ is either a parallelogram or the four points are on a line (Figure 6). In both cases, the midpoints of AD and BC coincide, and D is the mirror image of A through this midpoint. In the special case where B and C are the same lattice point, then D is the mirror image of A with respect to that lattice point (Figure 6c). We summarize our results here:

Theorem 2 (a) *In a point lattice, a line either contains at most one lattice point or infinitely many evenly spaced lattice points.*

(b) *If through every lattice point we draw a line parallel to a given lattice line, then each line is a lattice line with infinitely many lattice points and the distance between any two neighboring lattice points is the same.*

(c) *The mirror image of a lattice point either through another lattice point or through the midpoint of a lattice interval is also a lattice point.*

6. The points with integer coordinates can be viewed as the lattice points of a square lattice. Using Theorem 2, we can answer the following question: Which regular polygons can occur as lattice polygons (in an appropriate lattice)?

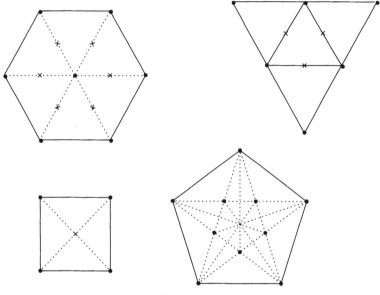

FIGURE **7.**

In a regular lattice polygon we reflect each vertex about the midpoint of the line segment adjoining its two neighboring vertices (Figure 7). By Part (c) of the theorem these are all lattice points again. In the case of a triangle, the new points determine a new equilateral triangle whose sides' midpoints are the vertices of the original triangle. In a square, each vertex is reflected into the opposite vertex; for a hexagon, all vertices are reflected into the center.

For polygons with a different number of sides, the reflected points would be the vertices of a smaller polygon with the same number of sides. This is not possible, since repeating this reflection process would lead to infinitely many lattice polygons, and hence infinitely many lattice points all in the interior of the first polygon. There would be arbitrarily small distances occurring among these points, contradicting property II.

7. Lattice squares can occur, as we have seen, simply by building a lattice out of squares. If we build a lattice out of rhombi with acute angle 60°, then we see that it contains lattice equilateral triangles and lattice regular hexagons, too (Figure 8). Regular lattice hexagons and lattice equilateral triangles occur together, for if we have a regular hexagon, taking every other vertex we have the vertices of an equilateral triangle. If ABC is a lattice equilateral triangle, then the six lattice points B, C, the mirror images of B through the midpoint of AC and through A, and the mirror images of C through A and the midpoint of AB are the vertices of a regular hexagon (Figure 8).

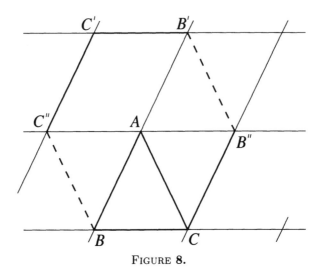

FIGURE 8.

We will show that if a lattice contains a lattice square, then it cannot contain a lattice equilateral triangle (nor a lattice regular hexagon). Choose for a coordinate system two adjacent sides of a lattice square, and let the unit length be the length of a side of the square. In this way the axes are lattice lines. The lattice lines parallel to the x-axis cut the y-axis into intervals of equal length, and distances between neighboring cut points are therefore rational. Every lattice point is on one of these lines, and therefore the y-coordinate is rational. For the same reason, the x-coordinate is also rational.

We conclude that the slope of any lattice line is also rational, and by the formula

$$\tan(\alpha - \beta) = \frac{\tan \alpha - \tan \beta}{1 - \tan \alpha \tan \beta}$$

we see that the tangent of the angle of intersection of two lattice lines is also rational. The angle $60°$ that occurs between sides of an equilateral triangle cannot occur, since $\tan 60° = \sqrt{3}$ is irrational. We summarize our results here:

Theorem 3 *In a parallelogram lattice, the only regular polygons that can occur as lattice polygons are triangles, squares, and hexagons.*

If there is a lattice square, then there is no lattice equilateral triangle nor a lattice regular hexagon. If there is either a lattice equilateral triangle or a lattice regular hexagon, then the other also occurs.

This theorem is due to T. GALLAI and P. TURÁN, who proved it as students. The proof presented for the first part is due to F. KÁRTESZI.[5]

8. We mentioned that a point lattice can be obtained from a parallelogram lattice in different ways. It is not difficult to determine all such parallelogram lattices.

We first note that a parallelogram lattice is determined by a single base parallelogram. More precisely, for any parallelogram there corresponds a (unique) parallelogram lattice for which the parallelogram is a base parallelogram. The pairs of opposite sides define neighboring lines in each of the two families, and the remaining lines in the families are parallel to these, at intervals whose distance is the distance between the corresponding pair of parallel sides. We say that *the parallelogram generates the parallelogram lattice.*

If a lattice figure does not contain lattice points other than its vertices, we call it *empty*. With these definitions we can now state the following theorem.

Theorem 4 *A point lattice is the point set of those parallelogram lattices that are generated by empty parallelograms.*

It is clear that if not every vertex of a parallelogram is a lattice point, or if there is a lattice point in the interior of the parallelogram or in the interior of one of its edges, then the point lattice it generates is not the same as the original one.

The important part of the theorem is that every empty parallelogram generates the point lattice. This is easily seen with the help of Theorem 2, Part (*b*). Let **P** be the given parallelogram lattice, **Q** its point lattice, and let *ABCD* be an empty parallelogram (Figure 9). Let **R** be the parallelogram lattice generated by *ABCD*, and call the point lattice it determines **S**. We need to show that **Q** and **S** are identical.

The lattice points on the lattice line *AB* of lattice **P** form a sequence of points at equal distances *AB*. According to the definition of **R**, its lattice threads parallel to *AD* cut the line just in these points.

These lattice threads of **R** are lattice lines occurring at a distance the same as that between *AD* and *BC*, and thus coincide with the lattice lines in **P** parallel to *AD*. On these lines, which can be viewed as either lattice lines of **P** or as the lattice threads of **R**, the lattice points occur at equal distances *AD*. In this way, **Q** and **S** coincide, finishing the proof.

---

[5] See *Mat. és Fiz. Lapok* **50.** (1943), pp. 182–183, problem 12 (in Hungarian).

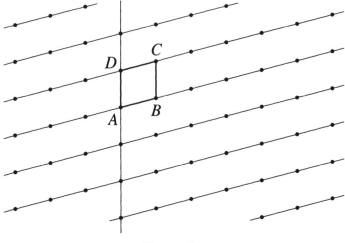

FIGURE 9.

9. Based on Theorem 4, it seems very likely that in a parallelogram lattice, all empty lattice parallelograms have the same area. It is an equivalent statement that all empty lattice triangles have the same area. The more general statement that we present as the next theorem is due to G. PICK.[6]

Theorem 5 *If the area of the base parallelograms in a parallelogram lattice is d, and a lattice polygon contains h lattice points in its interior and b lattice points on its boundary, then its area T is*

$$T = \left(b + \frac{h}{2} - 1\right) d.$$

This shows that the area of the polygon is determined only by the number of lattice points, and not by its shape. Every empty lattice quadrilateral, for example, has area d ($b = 0$ and $h = 4$); and every empty lattice triangle has area $d/2$.

The theorem is clearly true for those parallelograms P whose sides lie on lattice threads: Translate the lattice so that the lattice points move to the midpoints of the base parallelograms (Figure 10, the dashed lines) and consider the lattice parallelograms arising from the translated lattice. Every lattice point in the interior of P is the center of a parallelogram of area d from the translated lattice. The lattice points of P on the boundary that are not vertices are surrounded by parallelograms of area $d/2$; the small parallelograms that arise around the vertices of P together have area d. Putting this together, we get

$$bd + \frac{(h-4)d}{2} + d = \left(b + \frac{h}{2} - 1\right) d$$

---

[6] G. PICK: *Lotos Prag.* (2) **19** (1900), pp. 311–319.

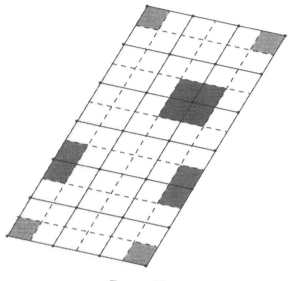

FIGURE **10.**

as the area, verifying the theorem in this case.

Denote the expression on right side of the equality in the theorem by K. The following provides the key to finishing the proof for all polygons:

Lemma *Let R be a lattice polygon; assume that R is divided into two parts R_1 and R_2 by lattice line segments going through its interior. Then the following equations hold with the parameters of the polygons given with the appropriate indices:*

$$T = T_1 + T_2, \qquad K = K_1 + K_2.$$

(See Figure 11.)

Only the second of these needs to be proved. Let s be the number of lattice points on the dashed line. Then

$$b = b_1 + b_2 + s - 2,$$

because the two endpoints of the dashed line are also points of the boundary of R. If we count the points on the boundary of the two smaller polygons, we get twice the points on the boundary of R, except the endpoints of the dashed line, which were already taken into account. Hence

$$h_1 + h_2 = 2s + h - 2; \quad \text{i.e.,} \quad h = h_1 + h_2 - 2s + 2.$$

We substitute this into the left-hand side of the equation, to get

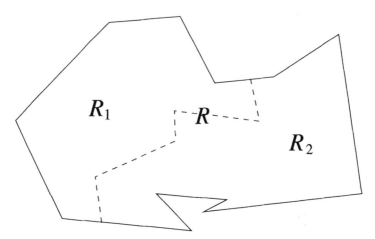

FIGURE 11.

$$K = \left(b_1 + b_2 + s - 2 + \frac{h_1 + h_2 - 2s + 2}{2} - 1 \right) d$$

$$= \left(b_1 + b_2 + \frac{h_1}{2} + \frac{h_2}{2} - 2 \right) d = K_1 + K_2.$$

With this we have proved the lemma.

We can rephrase the lemma as follows: If the theorem is true for two subpolygons of a polygon, then it is true for the polygon itself. This is of course true for more than two parts as well. We can also rephrase the lemma as follows: If the theorem is true for a polygon and a subpolygon of it, then it is also true for the other subpolygon.

10. We will now prove the theorem for triangles. We first verify it for triangles that have lattice threads as two sides. If AB and AC are lattice intervals of lattice threads through the lattice point A, and D is the mirror image of A through the midpoint of BC (Figure 12), then we have a parallelogram $ABDC$ for which the theorem is true. By Theorem 2, Part (c), the triangles CAB and BDC have the same area and contain the same number of lattice points. Thus $T_1 = T_2$ and $K_1 = K_2$, and

$$T = T_1 + T_2 = 2T_1 = K_1 + K_2 = 2K_1, \qquad T_1 = K_1,$$

showing that the theorem is true for the triangle ABC as well.

From this it already follows that the theorem is true for all lattice triangles. Given an arbitrary lattice triangle, we draw the lattice threads through the three vertices; we have two sets of three parallel lines (Figure 13). It is possible that some of these lines coincide. Among these lines, four determine a parallelogram enclosing the triangle. The lines are lattice threads, so the intersections are lattice points; hence the parallelogram is a lattice parallelogram.

Among the three vertices of the triangle, either all three are on the boundary of the parallelogram, or one side is a diagonal of the parallelogram, the two vertices are vertices of the parallelogram, and the third point is in the interior of the parallelogram. In the latter case, the lattice threads through this third vertex cut out a lattice parallelogram smaller than the original one.

We can then get the triangle by starting with the large parallelogram, removing the aforementioned parallelogram, if it is present, removing those triangles that have lattice threads as two of the sides (there can be three, two, or possibly just one of these). The theorem holds for the original parallelogram and all the parts we have removed from it. In this way, the theorem holds at each step for the remaining polygon; hence it is true for triangle ABC too.

We mention here that this shows that the area of all empty lattice triangles is the same, $d/2$, and that all empty lattice parallelograms have the same area d, since they can be split by a diagonal into two empty lattice triangles.

11. We now finish the proof for all lattice polygons[7] by induction on the number of vertices. We use the fact that every polygon with at least 4 sides

---

[7] We consider only simple polygons, i.e., those for which two sides do not intersect each other and vertices are contained on only two sides.

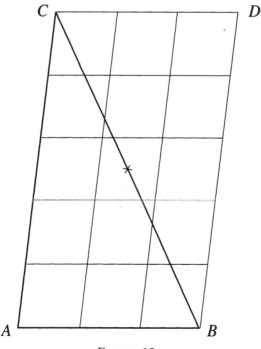

FIGURE **12.**

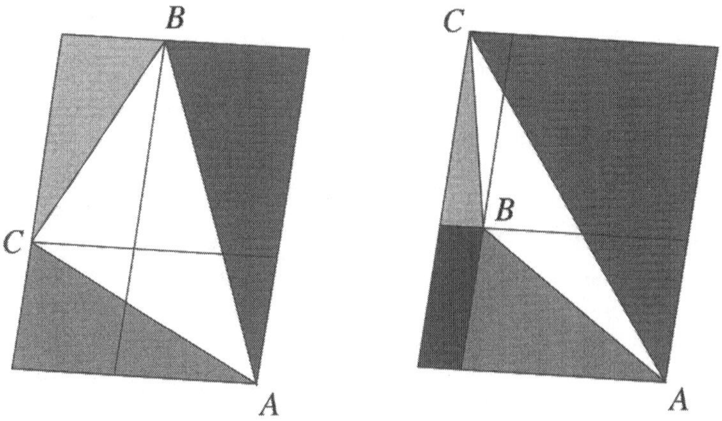

FIGURE **13.**

has a diagonal contained in its interior.[8] We will prove this last statement at the end.

We already know that Theorem 5 is true for all triangles. Assume that it is true for all polygons with fewer than n sides ($n > 3$). Let us examine such an n-gon. Let KL be a diagonal contained in the interior. Since it is a diagonal, both parts of the boundary between K and L contain additional points, and hence each of the two parts has fewer than n vertices. Thus the diagonal divides the figure into two polygons each with fewer than n vertices. By the assumption, the theorem is true for these two polygons, and by the lemma, it is then true for the original one too.

With this we have proved the theorem for all lattice polygons.

12. We now return to the proof of the existence of an interior diagonal. In a convex polygon, every diagonal is an interior diagonal. If the polygon is not convex, we consider its convex hull; this is also a polygon, whose vertices are all vertices of the original polygon having interior angle less than 180°. Let B be such a vertex on sides AB and BC of the convex hull. If there are no more vertices in the closed triangle ABC, with the exception of possible points on the intervals AB and BC, then the diagonal BC is in the interior of the polygon.

If this is not the case, then the triangle contains vertices of the polygon. Let D be the vertex or one of the vertices the furthest from AC contained by the triangle. (This can be on the line AC as well. See Figure 14.) Then the interval BD does not contain any vertices and is a diagonal satisfying the required condition.

[8] This theorem is of course true for all polygons, and has nothing to do with lattices.

Numerous proofs have been given for Pick's theorem (Theorem 5). This proof of the theorem is due to GY. PÓLYA.[9]

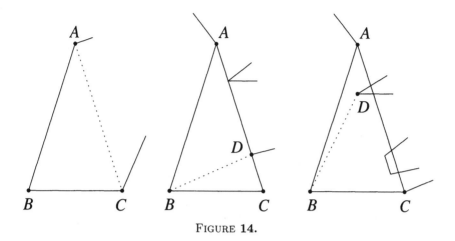

FIGURE 14.

Exercises:

5. Prove that if a lattice parallelogram contains an odd number of lattice points, then its center is a lattice point.

6. Prove that if a lattice triangle has no lattice points on its boundary in addition to its vertices, and one point in its interior, then this interior point is its center of gravity.

7. Prove that if a lattice parallelogram contains at most three lattice points in addition to its vertices, then those are on one of the diagonals.

8. Prove that if a point lattice (in the plane) contains a lattice square, then for every lattice line we can find a lattice line perpendicular to it.

    Does the same conclusion hold if we assume only that the lattice contains a lattice rectangle?

9. Prove that in every convex quadrilateral there is a vertex V such that if we complete the two sides from V to a parallelogram, then this parallelogram is contained in the quadrilateral.

    We note that more generally, it is also true that in any convex n-gon, there are at least $n - 3$ vertices for which each of the parallelograms

---

[9] G. L. ALEXANDERSON, J. PEDERSEN: *The Oregon Math. Teacher*, 1985. The presented format of the proof was given by ÁRPÁD SOMOGYI, who found it independently of PÓLYA.

obtained by completing the adjacent sides of the vertex is contained in the polygon.[10]

10. Prove that the only possible convex empty lattice polygons are triangles and parallelograms.

11*. Prove that in every parallelogram lattice there is an empty lattice triangle that has no angles greater than 90°, and all such triangles are congruent.

12. In the lattice of points with integer coordinates, consider a lattice rectangle with sides parallel to the coordinate axes, subdivided into empty lattice triangles. Prove that no matter how we subdivide the rectangle, we can always find in it a parallelogram together with one of its diagonals.

13. (a) Prove that in the subdivision of the previous exercise there always occurs half a square (two sides plus the diagonal).

(b*) Prove that the number of half-squares is at least twice the length of the shortest side.

14*. In the lattice of points with integer coordinates, let $PQRS$ be a lattice quadrilateral and E the intersection point of its diagonals. Prove that if the sum of the angles at P and Q is less than 180°, then the triangle PQE contains lattice points either in its interior or on its boundary other than P and Q (L. Lovász).

15. Prove Theorem 3 using Theorem 3.5 (Gallai–Turán).

16*. How close together do trees need to grow in a regularly planted circular forest so as to obstruct the view from the center? We rephrase this in the following problem:

Let s be a positive integer. In a lattice of unit squares we draw a circle of radius r around every lattice point contained in a circle of radius s centered at the origin. If r is relatively small, then there are rays from the origin that do not intersect any of the circles (we can see out through the forest). If r is large enough, then every ray from the origin has a common point with some circle. Prove that the critical value for r distinguishing the two cases is $r = 1/(s^2 + 1)$ (Gy. Pólya).

As we did for lattices in the plane, we can define lattices in (3-dimensional) space. Start with a point and passing through it three planes that intersect in three different lines. From these three planes we consider three families of planes; within each family the planes are parallel to the original one and equally spaced. The intersecting planes determine congruent parallelepipeds

---

[10] This problem appeared on the Miklós Schweizer Memorial Competition in 1964. This is a national mathematics competition in Hungary for high school students. See *Matematikai Lapok* **16** (1965), pp. 92–113, Problem 4 (in Hungarian) and Contests in Higher Mathematics, G. J. SZÉKELY (ed.), Springer, 1996, pp. 5 and 247–249.

with the same orientation. We call this a parallelepiped lattice. The vertices are the lattice points, and all the lattice points together are called the point lattice.

17. Rephrase Theorems 1–4 and Exercises 5–7 in terms of 3-dimensional lattices and determine which of them hold and which do not.

18. Do all empty lattice tetrahedra have equal volume?

19*. Prove that in a 3-dimensional lattice, all empty, convex, quadrangle-based pyramids have equal volume. Is convexity important here?

13. Returning to lattices in the plane, if we consider two parallel neighboring lattice lines, then a line parallel to these and falling between them cannot contain lattice points; in fact, there is a positive number such that no lattice point is closer than this distance to the line. If a line is not parallel to a lattice line, it is still possible that it contains a lattice point, but it may also contain none. We prove, however, the following fact:

Theorem 6 *If a straight line is not parallel to a lattice line, then on each side of the line there are lattice points arbitrarily close to the line.*

Let e be a line that is not parallel to a lattice line. Let A and B be lattice points, one on each side of e, and draw the lines a and b parallel to e through A and B, respectively (Figure 15). These lines do not contain more lattice points. Let d be the distance between a and b. The interval of a lattice line between a and b, parallel to the lattice lines through A and B, has the same length as the interval AB, and its endpoints on a and b are not lattice points; hence there is a lattice point C on this interval between a and b. Call this point C and let c be the line through it parallel to e.

The line c divides the strip between a and b into two smaller strips, one of which has width less than $d/2$. (It is not possible that it is equally distant from the two lines. Why?) We may assume that cc is closer to a. The two points on the lattice line through A and C that are closest to e, one on each side, or if a lattice point falls on e, then the two neighboring points, are lattice points at a distance at most $d/2$ from e.

Repeating this process with the two new points in place of A and B gives newer and newer points. The newer points are at distances at most $d/4$, $d/8$, etc. from e. With this we have proved the statement of the theorem.

14. We sharpen the previous result a little:

Theorem 6′ *The claim of Theorem 6 holds for rays not parallel to lattice lines. Further, for arbitrary positive h, there are infinitely many lattice points on both sides of the ray at a distance less than h from the ray.*

The previous proof can be applied with a minor adjustment. We restrict ourselves to the half-plane containing the ray, bounded by the line through

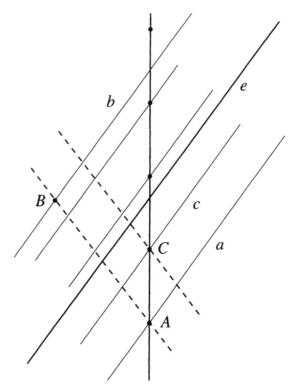

FIGURE 15.

the endpoint of the ray and perpendicular to it (Figure 16). In this half-plane we choose two lattice points A and B, one on each side of the ray. We then consider a lattice line parallel to AB on the side of AB not containing the endpoint of the ray, and we choose C on this. (In this figure, we have assumed C is closer to b then to a.) With this modification, we can find points on both sides of the ray arbitrarily close to it. This proves the theorem.

15. With the help of the previous theorem, we will prove the following statement, which in itself is not at all self-evident:

Theorem 7 *If b is a positive integer that is not a power of 10, then for any finite sequence of digits, there is a power of b whose leading digits are the same as the given sequence.*

Let n be the number of digits and A the number with the given sequence of digits. The fact that b^k starts with the given sequence of digits means that there is a number $m, m \geq n$, satisfying the two inequalities

$$A \cdot 10^m \leq b^k < (A+1)10^m.$$

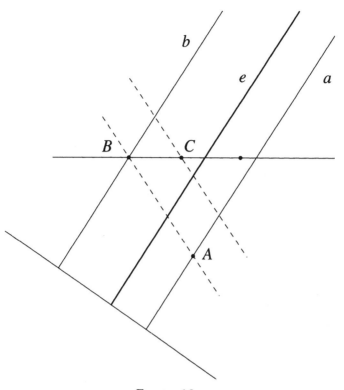

FIGURE **16.**

The base-10 logarithm is a monotone increasing function, so we can obtain equivalent statements using logarithms. The statements we get in this way can be rearranged as

$$k \log_{10} b - \log_{10}(A + 1) < m \leq k \log_{10} b - \log_{10} A.$$

The two outer expressions are first-degree functions of k. For positive k, these are represented by rays in the coordinate plane, both having slope $\log_{10} b$, with endpoints $(0, -\log_{10}(A+1))$ and $(0, -\log_{10} A)$ (Figure 17). Since m is a nonnegative integer, we are looking for an integer lattice point (k, ℓ) that is in the strip between the two lines in the first quadrant. Lattice points below the upper line and at a distance from it less than the distance between the two lines will suffice.

To show that such points exist, it is enough to show that the two bordering lines of the strip are not parallel to lattice lines.

In the lattice of points with integer coordinates, all lattice lines have rational slope. It is therefore enough to show that $\log_{10} b$ is irrational. Assume that it is rational, meaning that $\log_{10} b = u/v$, or that $b^v = 10^u = 2^u 5^u$, where u and v are integers. By the fundamental theorem, b cannot have any divisors other than 2 and 5, so we may write $b = 2^r 5^s$. Substituting this into

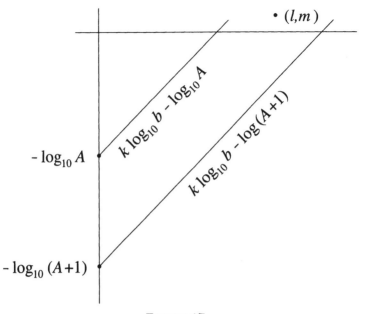

FIGURE 17.

the previous equation, and using the fundamental theorem, we get

$$rv = u = sv, \quad \text{and therefore} \quad r = s, \quad b = 2^r 5^r = 10^r.$$

This was excluded at the beginning; hence we have proved the theorem.

Exercises:

20. Prove that if α is an irrational number, then the numbers $\{n\alpha\} = (n\alpha - [n\alpha])$ $(n = 1, 2, \dots)$ are dense in the interval $(0, 1)$. By *dense* we mean that for every positive $\delta > 0$ and $0 < \beta < 1$, there is an n such that

$$|\{n\alpha\} - \beta| < \delta.$$

21. Prove that if b is an integer larger than 1, then the number obtained by writing the powers of b in order after the decimal point is an irrational number. (For example, in the case of $b = 6$, we get the number

$$0.63621612967776465656\dots.)$$

16. We arrived at parallelogram lattices by interpreting geometrically questions concerning Diophantine approximation. We saw (using a different geometric interpretation in Theorems 3.7 and 3.8) that for a real number α and a positive integer q, there are integers u and v such that

$$|v| \leq q, \quad \text{and} \quad |v\alpha - u| \leq \frac{1}{q+1}.$$

Here we interpret the left-hand sides of these as absolute values of the ordinate and abscissa of a point. Then the inequalities

$$|x| < \frac{1}{q+1}, \qquad |y| < q$$

(with arbitrary u and v) are satisfied for the coordinates of the points of a rectangle with sides parallel to the axes and centered about the origin. We can hope to solve some other questions in Diophantine approximation by giving theorems on the existence of lattice points in regions under certain conditions. The fundamental result in this direction is the following:

Theorem 8 (Minkowski's Theorem) (a) *In a point lattice in which empty parallelograms have area d, every convex, centrally symmetric region of area greater than $4d$, centered at a lattice point, contains a lattice point in its interior in addition to its center.*

(b) *If the area of the region is exactly $4d$, then in addition to its center, there is either a lattice point on its boundary or in its interior.*

MINKOWSKI's result is more general, valid for arbitrary dimension.

17. The following simple proof follows from a fundamental idea that was often forgotten and rediscovered.[11] Let L be a lattice in the plane with empty lattice parallelograms having area d. Let T be a centrally symmetric convex region of area greater than $4d$ in the plane centered at a lattice point O.

Starting from an empty lattice parallelogram $OABC$, take every other lattice thread parallel to the lines through OA and OC. In this way we cover the plane by parallelograms having area $4d$. Let $KLMN$ be an arbitrary parallelogram from these, and translate into $KLMN$ those other parallelograms that intersect T, along with the part of T that they contain (Figure 18). In this way the regions we have translated have together more than $4d$ area and are all in a parallelogram of area $4d$; hence there is a point P covered by at least two translated regions.

Fix the points of two regions covering P and retranslate them back to their original position in such a way that at every step of the translation we move the regions by a distance equal to one side of the parallelogram in a direction parallel to that side. Call P_1 and P_2 the two points whose images are P. Putting together the two (re)translations, we get a path from P_1 to P_2 that is made up of intervals parallel to OA or OC and twice their length.

[11] G. BIRKHOFF (1914), referred to in H. F. BLICHFELDT: *Transactions Amer. Math. Soc.* **15** (1914), pp. 227–235. W. SCHERRER: *Math. Annalen* **89** (1923), pp. 255–259, and *Dissertation*, Universität Zürich, 1923. GY. HAJÓS: *Acta Sci. Math. Szeged,* **6** (1934), pp. 224–225.

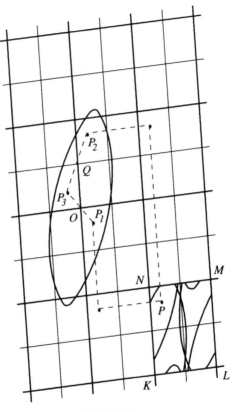

FIGURE **18.**

Let P_3 be the mirror image of P_1 with respect to O. This is also a point of T, since T is centrally symmetric. Shrink the path from P_3 in half. The image of P_1 by this shrinking is O, and let Q be the image of P_2. This is also a point of T, because by convexity T contains the interval P_2P_3. This is a lattice point, since there is a path from O to Q that has intervals parallel to and equal in length to OA and OC, respectively. Finally, Q is different from O, since they are the images of the distinct points P_1 and P_2 under the shrinking. With this we have proved Minkowski's theorem in the case where the area of T is greater than $4d$.

18. This procedure can also be used without any changes when the area of the parallelogram is exactly $4d$, if the parallelogram $KLMN$ still has a point that is covered by more than one translated region.

However, if this is not the case, then the relocated pieces (including their boundaries) fill the parallelogram. This contains 9 lattice points. Of those, O gets mapped to the four vertices when we shift the parallelograms meeting at O onto $KLMN$, so O cannot have any other image.

The remaining 5 lattice points are also in some of the translated regions, or on their boundaries. Then the piece that was moved there includes a lattice point different from O, which is inside the parallelogram or is on its boundary, and hence the second claim of Minkowski's theorem is true.

19. We do indeed need the assumptions of the theorem. The area of $4d$ cannot be decreased: For instance, in the lattice of points with integer coordinates the square with vertices $(\pm 1, \pm 1)$ contains only lattice points on its boundary apart from the origin, and if we shrink the region (keeping the origin fixed) by however small an amount, the only lattice point belonging to the square will be the origin, although its area can be arbitrarily close to $4d$.

We cannot give up convexity: The figure that is the union of the square having vertices $\left(\pm\frac{3}{4}, \pm\frac{3}{4}\right)$ and the two rectangles with area T whose points have coordinates x and y satisfying

$$\frac{1}{4} \le |x| \le \frac{3}{4}, \qquad |y| \le T, \qquad xy > 0$$

(Figure 19), is symmetric about the origin, its area (depending on T) can be as large as we want, and the only lattice point contained by the region is the origin.

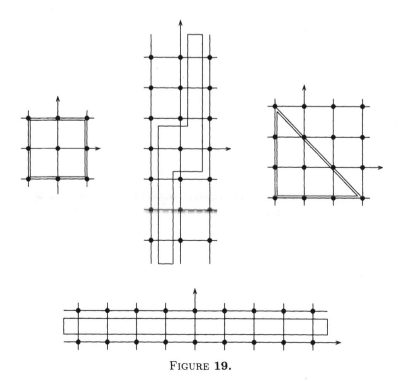

FIGURE **19.**

Central symmetry cannot be ignored either: the area of the triangle with vertices $(-1, -1), (-1, 2)$, and $(2, -1)$ is $\frac{9}{2}$, but if we shrink its size by a very small amount, it will not contain any lattice points other than the origin (its center of gravity), yet its area is still larger than $4d$. The rectangle with points bounded by the inequalities

$$-T \leq x \leq T, \qquad \frac{1}{4} \leq y \leq \frac{3}{4},$$

can have an arbitrarily large area (depending on T), and again, it does not contain a lattice point.

20. The center of gravity of a region that is central symmetric is at its center. From this, the following question arises: "What can be said about convex regions whose center of gravity is a lattice point?" E. ERHART showed that *if in a lattice the area of an empty parallelogram is d, then all convex regions with center of gravity a lattice point and area at least 9d/4 contain at least one additional lattice point.*[12] The above example with the triangle shows that we cannot improve the $9d/4$ limit. We do not have a corresponding sharp theorem in higher dimensions.

21. As an application of the above we prove the following theorems.

Theorem 9 *In a planar lattice for which the origin is a lattice point and the area of the base parallelograms is d, if k_1 and k_2 are positive numbers whose product is not less than d, then there exists a lattice point different from the origin whose coordinates satisfy*

$$|x| \leq k_1, \qquad |y| \leq k_2. \tag{3}$$

Theorem 10 *With the conditions of the previous theorem, for any arbitrary positive number c, there exists a lattice point other than the origin for which*

$$c|x| + \frac{1}{c}|y| \leq \sqrt{2d}. \tag{4}$$

Theorem 11 *Using the notation of Theorem 9, there exists a lattice point differing from the origin for which the following inequality holds:*

$$x^2 + y^2 \leq \frac{4d}{\pi}. \tag{5}$$

All three statements are direct consequences of Minkowski's theorem. The conditions of (3) are satisfied for a parallelogram that is symmetric about the origin and whose sides are parallel to the axes, with lengths $2k_1, 2k_2$; i.e., its area is at least $4d$.

---

[12] E. ERHART: *Comptes Rendus Acad. Sci. Paris* **240** (1955), pp. 483–485.

The condition in (4) holds for a rhombus with vertices $\left(\pm\sqrt{2d}/c, 0\right)$ and $\left(0, \pm c\sqrt{2d}\right)$. The area of this is also $4d$.

Finally, (5) describes the circle centered at the origin with radius $2\sqrt{d/\pi}$, whose area is again $4d$. Hence by Minkowski's theorem, all of them contain (inside the region or on its boundary) a lattice point other than the origin.

22. Using Theorems 9 and 10, we can draw conclusions on the approximation of real numbers by rational numbers. At the beginning of this chapter we saw that if α is a real number, then we should examine the lattice with points having coordinates satisfying

$$x = -u + v\alpha, \qquad y = v, \tag{6}$$

where u and v are integers. For this lattice, $d = 1$. For Theorem 9, then, $k_1 = 1/k_2$ is an appropriate choice. This is really Theorem 3.7 with a slight alteration.

Theorem 10 gives a better approximation. If we were simply to ignore the second summand of the left-hand side of (4), then we would conclude only that for any large positive c we can find integers u and v satisfying $|\alpha - u/v| \leq \sqrt{2d}/cv$. (We may assume that v is positive, since in addition to (u, v), the pair $(-u, -v)$ also satisfies the hypotheses of Part (b).)

We get a much better result if we apply the inequality between the geometric and arithmetic means of the left-hand side. This gives, at the moment for an arbitrary lattice,

$$|xy| \leq \left(\frac{c|x| + 1/c \cdot |y|}{2}\right)^2 \leq \frac{d}{2}.$$

Applying this to the lattice (6) and dividing by v^2 we get that for appropriate integers u and v,

$$\left|\alpha - \frac{u}{v}\right| \leq \frac{1}{2v^2}, \tag{7}$$

which is a better approximation than the previous one.

Apparently, the factor c does not play a role here, but this will change when we prove a theorem similar to Theorem 3.8. If we suppose that α is irrational, then it helps to prove that there are infinitely many fractions for which (7) holds.

First we prove the following for an arbitrary lattice:

Theorem 12 *In a planar lattice for which the area of empty parallelograms is d, there are infinitely many lattice points whose coordinates satisfy*

$$|xy| < \frac{d}{2}.$$

Notice that this inequality holds for points between the two branches of a hyperbola, which is not a convex region, but the rhombus corresponding to Theorem 10 is part of this region.

We also note that the inequality we wrote above is strict. That is because we used the inequality for algebraic and geometric means, and between these two means equality holds only if the two numbers are equal. In our case, this means

$$c|x| = \frac{1}{c}|y|.$$

This, however, describes the midpoints of the sides of the rhombus. Therefore, these have to lie on the hyperbolas (the boundary of the region), and the rest of the sides of the rhombus must lie between the hyperbolas. This means that the sides of the rhombus are tangent to the hyperbolas.

So even if there are lattice points on the hyperbolas, we can choose the rhombus, i.e., the value of c, such that its sides do not touch the hyperbolas in lattice points.

23. Returning to the proof of the theorem, if one of the axes has a lattice point other than the origin, then it has infinitely many of them, and for these the inequality of the theorem holds. If, for instance, the only lattice point on the y-axis is the origin itself, then we already know that there exists an (x_0, y_0) lattice point inside the region bounded by the hyperbolas. Then we apply Theorem 10 with c_1 greater than $\sqrt{2d}/|x_0|$ (and for which the sides of the rhombus are not tangent to the hyperbolas at lattice points; see Figure 20). This yields a lattice point (x_1, y_1) for which

$$c_1|x_1| < \sqrt{2d}, \qquad \text{i.e.,} \qquad |x_1| < \frac{\sqrt{2d}}{c_1} < \frac{\sqrt{2d}}{\sqrt{2d}/|x_0|} = |x_0|.$$

Thus it is different from the previous lattice point.

Since there are not lattice points on the y-axis, $x_1 \neq 0$, and this way the procedure can be repeated as many times as we want. This proves the claim of the theorem.

Applying Theorem 12 specifically to lattices of type (6), we obtain the following theorem.

Theorem 12′ *For every irrational number α there are infinitely many fractions u_i/v_i for which*

$$\left| \alpha - \frac{u_i}{v_i} \right| < \frac{1}{2v_i^2}.$$

We will soon see that the approximation can be further improved upon.

24. One of the fundamental properties of lattices, which we discussed in the introduction of this section, is that the distances between lattice points cannot be arbitrarily small. Theorem 11 gives an upper bound for that smallest distance. It can be used to prove the following number-theoretic theorem:

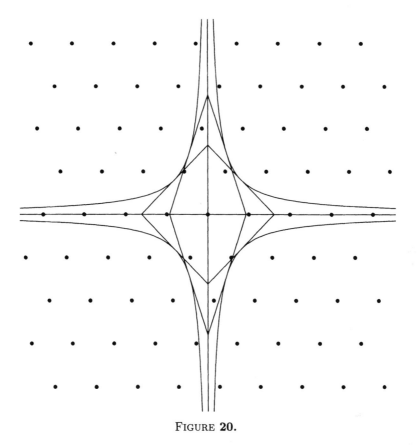

FIGURE **20.**

Theorem 13 *The prime 2 and positive primes of the form $4k + 1$ can be written as the sum of the squares of two positive integers. Numbers of the form $4k + 3$ cannot be written in such a way.*

The last claim is obvious. Squares of even numbers are divisible by 4, and squares of odd numbers have remainder 1 upon division by 4 (in fact, they have remainder 1 upon division by 8 by Theorem 1.14). Hence, the sum of two squares can have remainder 0, 1, or 2, but not 3, upon division by 4.

By Theorem 2.15, for primes p of the form $4k + 1$, there exists an integer a for which $a^2 + 1$ is divisible by p (for $p = 2$ taking $a = 1$ suffices). For such an a, let us examine the lattice described by

$$x = pu + av, \qquad y = v,$$

where u and v are integers. For this lattice, $d = p$, and the sum of the squares of coordinates of lattice points is divisible by p, since

$$x^2 + y^2 = p\left(pu^2 + 2auv\right) + \left(a^2 + 1\right)v^2,$$

which is divisible by p by the choice of a.

By Theorem 11, there is a lattice point whose coordinates satisfy

$$x^2 + y^2 < \frac{4p}{\pi} < 2p.$$

Since this lattice point is not the origin, the sum $x^2 + y^2$ is positive, and we have already shown that it is divisible by p. This is possible only if $x^2 + y^2 = p$. Since x and y are integers, this finishes the proof of the theorem.[13]

Exercises:

In the exercises of Section 12 we defined the parallelepiped lattice. The following exercises use those definitions.

22. Prove that in a parallelepiped lattice in which the volume of an empty parallelepiped is d and the origin is a lattice point, if the volume of a convex region, symmetric with respect to the origin, is greater than $8d$, then that region has a lattice point in its interior other than the origin. If the volume is exactly $8d$, then there is a lattice point either inside the region or on its boundary.

23. (Continuation) Let

$$x = a_1 u + b_1 v + c_1 w, \quad y = a_2 u + b_2 v + c_2 w, \quad z = a_3 u + b_3 v + c_3 w,$$

$$d = \left| \det \begin{pmatrix} a_1 & b_1 & c_1 \\ a_2 & b_2 & c_2 \\ a_3 & b_3 & c_3 \end{pmatrix} \right| > 0.$$

State and prove the three-dimensional generalizations of Theorems 9, 10, and 11.

24*. Draw a sphere around one lattice point of a three-dimensional lattice such that the sphere goes through the lattice points nearest to it. Prove that the lattice can be modified (if it does not already share this property) so that the volume of the empty parallelepipeds decreases and 6 coplanar lattice points are moved onto the surface of the sphere.

25. (Continuation) Using the notation of Exercise 23, prove that there is a triple of integers (u, v, w) other than $(0, 0, 0)$ for which

$$x^2 + y^2 + z^2 \leq \sqrt[3]{2d^2},$$

and this bound cannot be improved.

26*. We would like to approximate the real numbers α and β by two fractions with the same denominator, u/w and v/w. What can be said about the sum of the squares of the differences between the real numbers and their approximations when it is given that $0 < w \leq q$ (q is a given positive number)?

---

[13] This proof was supplied by P. TURÁN, who adapted an idea of HERMITE, who proved a related theorem about the sum of four squares (as we will see in Theorem 7.4'); See H. DAVENPORT: *Math. Gazette* **31** (1947), pp. 206–210.

27*. For what values of c are there integer solutions u, v, w ($w \neq 0$) to the inequality

$$\left| w \left((\alpha w - u)^2 + (\beta w - v)^2 \right) \right| \leq c$$

for arbitrary real numbers α and β? Compare the answer to that of the previous exercise.

We note that for Exercise 26, MAHLER[14] determined the smallest bound for which the claim is still valid; for Exercise 27, the smallest c was determined by DAVENPORT and MAHLER.[15]

25. Theorems 9 through 12 (and Minkowski's theorem as well) can be rephrased by fixing the region: Given a convex, centrally symmetric region of area $4d$, then every lattice with a lattice point at the center of the region whose empty parallelograms have area less than d has another lattice point inside the given region.

We mentioned when we derived Theorem 12′ from Theorem 12 that the estimation we obtained can be improved. For given k_1 and k_2 in Theorem 9 equality holds only for the lattice generated by the rectangle with vertices $(0,0), (k_1, 0), (0, k_2)$, and (k_1, k_2). The area of the rectangle is obviously $d = k_1 k_2$. If we enlarge this lattice by however small a factor, the lattice will not have a lattice point inside nor on the perimeter of the polygon other than the origin. Hence the inequality cannot be sharpened.

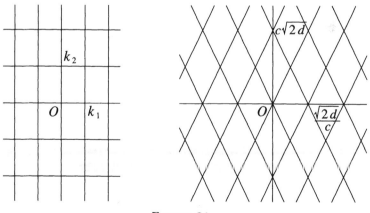

FIGURE **21.**

Similarly, no improvements can be made in Theorem 10, since the lattice generated by the rhombus

[14] K. MAHLER: *Quarterly Journal of Math., Oxford Ser.* **17** (1946), pp. 16–18.
[15] H. DAVENPORT, K. MAHLER: *Duke Math. Journ.* **13** (1946), pp. 105–111.

$$(0,0), \quad \left(\frac{\sqrt{d/2}}{c}, \pm\left(\sqrt{d/2}\right)c\right), \quad \left(\sqrt{2dc}, 0\right)$$

has lattice points only on the perimeter of the big rhombus, other than the origin, and the area of the small rhombus is d (Figure 21).

26. The inequality in Theorem 11 holds for a disk of radius $r = 2\sqrt{d/\pi}$. In this case there exists a lattice whose lattice points (other than the origin) occur only on the perimeter of the disk, forming the vertices of a regular hexagon. Such a lattice can be generated by a rhombus with sides r and an angle of $60°$. The area of such a rhombus is

$$d' = \frac{r^2\sqrt{3}}{2} = \frac{2\sqrt{3}}{\pi}d > d.$$

It is expected, but not at all obvious, that all lattices generated by parallelograms of area smaller than d' have lattice points inside the above disk. We will indeed prove that this is the case.

Theorem 14 *Every lattice in which the area of empty lattice parallelograms is not greater than $r^2\sqrt{3}/2$ has in addition to the origin a lattice point belonging to the disk*

$$x^2 + y^2 \le r^2.$$

Theorem 11, in contrast, provides only the value $r^2\pi/4$ in place of $r^2\sqrt{3}/2$. In comparison, $\sqrt{3}/2$ is approximately 0.866025, while $\pi/4$ is approximately 0.785398.

This problem can also be viewed as an algebraic question. The lattice points can be written as

$$x = au + bv, \quad y = a'u + b'v,$$

where a, b, a', and b' are given real numbers,

$$d = |ab' - a'b|;$$

u and v can be arbitrary integers. The question now is, how small can the sum of the squares of two homogeneous, linear binary expressions be for pairs of integers (x, y) other than $(0,0)$?

This question has also been examined for the sum of squares of linear forms with more than two variables. The exact limits are known up to the sum of squares of eight forms with eight variables. Proofs have been given for the cases 9 and 10, but these are incomplete.[16]

---

[16] See T. W. CHAUNDY: *Quarterly Journal of Mathematics, Oxford Ser.* **17** (67) (1946), pp. 166–192. The statement of Theorem 13 there is due to GAUSS.

27. To simplify the arguments in the proof of the above theorem, we call a lattice *admissible* for a region T, or T-*admissible*, when it does not have lattice points (other than the origin) inside the region T.

Denote the disk $x^2 + y^2 \leq r^2$ by K and let the area of the empty parallelograms in a K-admissible lattice be d. We will transform the lattice step by step so that more and more lattice points will be on the perimeter of K, the area of the empty parallelograms should decrease, or at least not get larger, and the lattice will stay K-admissible.

According to Minkowski's theorem, this lattice already has a point inside the disk of radius $\sqrt{2d}$, and by property II, there cannot be infinitely many of these points inside the disk. With an appropriate contraction of the lattice centered at the origin we can move an opposing pair of lattice points P and P' onto the perimeter of K, if there were none.

This transformation takes lattice points on lattice lines into points of equal distance between them (according to the proportion of contraction); it also transforms parallel lines into parallel lines, and points of intersection into points of intersection. Hence, in the new transformed lattice, the area of empty parallelograms is decreased by the square of the proportion of contraction.

In the following let us choose for the x-axis the line PP', without changing the origin. This way the inequality describing K does not change.

We modify the lattice (unless it originally has this property) such that the line PP' does not change, but the first lattice line parallel to and above PP' is shifted so that the lattice points intercepting the y-axis are moved to be at equal distances from this axis (Figure 22). This is achieved by a transformation of the form $x' = x + cy, y' = y$ for a suitable real value c (this is called a *shear transformation*). The images of the lattice points will be

$$x' = (a + ca')u + (b + cb')v, \qquad y' = a'u + b'v,$$

for integers u and v. These also form a lattice. The area of the new empty lattice parallelograms is

$$d' = |(a + ca')b' - a'(b + cb')| = |ab' - a'b| = d,$$

i.e., it has not changed.

Geometrically, the transformation leaves the x-axis unchanged; lines parallel to it get transformed into themselves; and lines through the origin are rotated around it. In this way, the side on the x-axis of a parallelogram that generated the lattice and its vertical height remain unchanged. Hence its area remains unchanged. The lattice is still admissible after the transformation, since one of the two lattice points mentioned above moved further away from the origin, hence still outside the disk if it was originally outside it. The other, by the established symmetry, is also outside the disk.

After this, we take this pair of lattice points (which are now symmetric with respect to the y-axis) and with a compression transformation (the

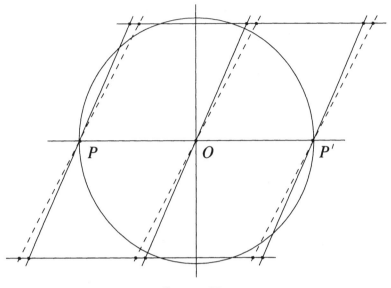

FIGURE **22.**

direction of the compression perpendicular to the x-axis) move them onto the perimeter of the disk, if they were not already there. This compression decreases the area of the empty parallelograms.

The K-inscribed hexagon that arises in this way is regular, since on the lattice lines parallel to PP', consecutive lattice points are at a distance r from each other. Hence, for every K-admissible lattice we have created another admissible lattice in which the area of the empty parallelograms is not greater than in the original, and six of its lattice points form a regular hexagon on the perimeter of K. The lattice can be generated by a rhombus of area $r^2\sqrt{3}/2$, and we have proved Theorem 14.

28. In Theorems 12 and 12′, the factor $\frac{1}{2}$ on the right-hand side of the inequalities can be replaced by $1/\sqrt{5}$. This yields the following theorem.

Theorem 15 *A lattice in which the area of the empty parallelograms is d has infinitely many lattice points whose coordinates satisfy the inequality*

$$|xy| \leq \frac{d}{\sqrt{5}}. \tag{8}$$

This theorem is due to KORKINE and ZOLOTAREFF.[17] Rephrasing the problem again in terms of linear forms, DAVENPORT determined the minimum value of the product of the values of three linear forms in three variables, taken

---

[17] A. KORKINE, G. ZOLOTAREFF: *Math. Annalen* **6** (1873), pp. 366–389.

at integer places.[18] For linear forms with more than 3 unknowns the exact value of the constant is not yet known.

In the proof of this theorem we can ignore again those lattices that have lattice points on the y-axis (in addition to the origin), since there would then be infinitely many of them, and for these $xy = 0$ holds, and hence (8) would be trivially satisfied for infinitely many points.

The line of thought in the previous proof is not applicable here. The reason is that there we started with an admissible lattice that already had lattice points, but not infinitely many in a region containing K, and with an appropriate contraction we were able to move the nearest lattice point onto the perimeter of K.

Here we are examining lattices admissible for the region between the pair of hyperbolas described by (8). This region is not bounded, so it is possible that there are lattice points arbitrarily close to the hyperbolas without actually being on the boundary of the region. It is therefore not possible to use a contraction to bring them onto the boundary. We need to find some other method. For this purpose, we will examine the structure of lattices a bit more closly.

29. Take a lattice point P in the first quadrant and call its orthogonal projection onto the x-axis Q. The triangle OPQ (where O is the origin) contains at most finitely many lattice points, and hence we can choose a lattice point P_0 for which the corresponding triangle OP_0Q_0 does not contain any other lattice points apart from O and P_0. Consider the lattice line parallel to and above the line OP_0. Let P_1 and P_2 be the lattice points on this line intercepting the y-axis from the left and right sides, respectively. Let x_i and y_i denote the absolute values of the coordinates of the points P_i, and M_i the product x_iy_i $(i = 0, 1, 2)$.

Since $OP_0P_2P_1$ is a parallelogram,

$$x_2 = x_0 - x_1 \qquad \text{and} \qquad y_2 = y_0 + y_1.$$

Based on this, the following relation holds for the M_i's:

$$M_2 = M_0 - M_1 + x_0y_1 - x_1y_0,$$

or rearranging terms,

$$-M_0 + M_1 + M_2 = x_0y_1 - x_1y_0. \qquad (9)$$

We can eliminate the coordinates, using the facts that the parallelogram is empty, and its area is d. In the following we will slightly change the shape

[18] H. DAVENPORT: *Proc. London Math. Soc.* **44** (1938), pp. 412–431, and *Journal London Math. Soc.* **16** (1941), pp. 98–101. For further results and problems see the books of J. W. S. CASSELS, and of P. M. GRUBER and C. G. LEKKERKERKER mentioned in footnote 4.

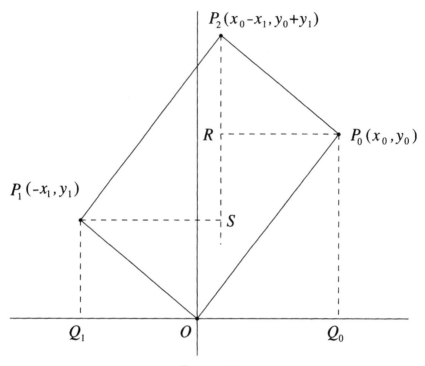

$P_2(x_0-x_1, y_0+y_1)$

R

$P_0(x_0, y_0)$

$P_1(-x_1, y_1)$

S

Q_1

O

Q_0

FIGURE **23.**

of the parallelogram (Figure 23). First we erect perpendiculars form each of the P_i's to the x-axis, and from P_0 and P_1 to the y-axis. Then, using the notation of the figure, we see that the area of the hexagon $Q_0P_0RSP_1Q_1$ is also d,

$$d = x_0y_1 + x_1y_0. \tag{10}$$

This is because of the congruences $\triangle P_0RP_2 \cong \triangle OP_1Q_1$ and $\triangle SP_2P_1 \cong \triangle Q_0P_0O$. Hence subtracting the squares of the sides of (9) from the squares of (10), we get

$$d^2 - (-M_0 + M_1 + M_2)^2 = 4x_0y_1x_1y_0 = 4M_0M_1,$$

or rearranging the terms, we have

$$d^2 = M_1(2M_0 + M_1 + 2M_2) + (M_2 - M_0)^2. \tag{11}$$

Denoting the smallest of the M_i's by M, we conclude that

$$d^2 \geq 5M^2, \quad \text{i.e.,} \quad M \leq \frac{d}{\sqrt{5}},$$

which means that (8) holds for the coordinates of at least one of P_0, P_1, and P_2. That is, there is at least one lattice point for which the claim of Theorem 15 is true.

Since we assumed that the origin is the only lattice point on the y-axis, by Theorem 6 there exists another lattice point within any arbitrary distance to this axis. We repeat the above procedure starting from a lattice point found closer than the previous one and in this way we get a new lattice point for which (8) holds. The procedure can be repeated over and over again, giving infinitely many lattice points satisfying (8), and hence the claim of Theorem 15.

30. Note that it is not possible to improve the bound, since if $M_1 = M_2 = M_3 = M$, then by (11) it follows that

$$d = M\sqrt{5} \quad \text{or} \quad M = \frac{d}{\sqrt{5}}.$$

From (9) and (10) we get

$$x_0 y_1 - x_1 y_0 = M = \frac{d}{\sqrt{5}}, \qquad x_0 y_1 + x_1 y_0 = d.$$

It then follows that

$$x_0 y_1 = \frac{\sqrt{5}+1}{2} \cdot \frac{d}{\sqrt{5}} \quad \text{and} \quad x_1 y_0 = \frac{\sqrt{5}-1}{2} \cdot \frac{d}{\sqrt{5}},$$

or by multiplying both sides by x_0 and y_0, respectively, and using the fact that $x_0 y_0 = M_0 = M = d/\sqrt{5}$, we have

$$y_1 = y_0 \frac{\sqrt{5}+1}{2} \quad \text{and} \quad x_1 = x_0 \frac{\sqrt{5}-1}{2}.$$

Moreover, OP_0 and OP_1 are the sides of an empty lattice parallelogram, and for the coordinates of an arbitrary lattice point, we get with appropriate integers u and v,

$$x = x_0 \left(u - \frac{\sqrt{5}-1}{2} v \right) \quad \text{and} \quad y = y_0 \left(u + \frac{\sqrt{5}+1}{2} v \right). \tag{12}$$

For these lattices,

$$d = x_0 y_0 \left(\frac{\sqrt{5}+1}{2} + \frac{\sqrt{5}-1}{2} \right) = x_0 y_0 \sqrt{5}$$

from the one side, and from the other side the absolute value of the product

$$xy = x_0 y_0 \left(u^2 + uv - v^2 \right)$$

is at least $x_0 y_0 = d/\sqrt{5}$, since $u^2 + uv - u^2$ is a nonzero integer when u and v are integers, not both 0.

So Theorem 15 is true with the following supplement.

Supplement. *In Theorem 15, equality holds only for those lattices described by (12). For other lattices, if at least one of the axes contains no lattice points (other than the origin), then the stronger relation*

$$0 < |xy| < \frac{d}{\sqrt{5}}$$

holds for coordinates of infinitely many lattice points.

31. We know that in lattices corresponding to Diophantine approximation the points on the x-axis with integer coordinates are lattice points. Lattices represented by (12) are not of this type, and hence Theorem 15 cannot immediately give the exact answer to this case. However, we will prove the following theorem:

Theorem 16 *For every irrational number α, there are infinitely many integers u and v satisfying*

$$\left| \alpha - \frac{u}{v} \right| < \frac{1}{\sqrt{5}v^2}.$$

If we replace $\sqrt{5}$ with a larger value, then there exist irrational numbers α for which the corresponding inequality is satisfied by at most finitely many fractions.

The first claim of Theorem 16 is a special case of the addition to Theorem 15 applied to the lattice described by the relations $x = u - \alpha v$ and $y = v$ (for u and v integers).

We can get to the proof of the second claim of Theorem 16 by analyzing the proof of Theorem 15. The infinitely many lattice points that we found in that proof come arbitrarily close to the y-axis, hence the absolute values of the abscissae get arbitrarily close to 0, while the absolute values of the ordinates grow without bound. Writing the coordinates of the lattice point in the form

$$x = au + bv, \qquad y = a'u + b'v,$$

we may assume that $ab' - a'b > 0$, so that its value is d.

Let us rewrite the ordinate in the following way:

$$y = \frac{a'}{a}(au + bv) + \frac{ab' - a'b}{a}v = \frac{a'}{a}x + \frac{d}{a}v.$$

Here the first term, together with x, becomes arbitrarily small. Hence, on the one hand, the area of the empty parallelograms in any lattice of the form

$$x' = \frac{1}{a}x = u + \frac{b}{a}v, \quad y' = dv = ay - a'x,$$

is still d, while on the other hand,

$$xy - x'y' = \frac{a'}{a}x^2$$

is also arbitrarily small given that x is small enough.

Applying our results to lattices of the form (12), we get that

$$x^* = u - \frac{\sqrt{5}-1}{2}v, \qquad y^* = dv \ \left(= x_0 y_0 \sqrt{5}v\right).$$

Since x^*, y^* are not of the form (12), $|x^*y^*| < d/\sqrt{5}$ holds for infinitely many integer pairs u, v, and in this way

$$\left|\left(u - \frac{\sqrt{5}-1}{2}v\right)v\right| < \frac{1}{\sqrt{5}}. \tag{13}$$

On the other hand, since in lattices of the form (12), the inequality

$$|xy| \geq \frac{d}{\sqrt{5}}$$

holds for lattice points other than the origin, it follows that $|x^*y^*|$ and with it the left-hand side of (13) get arbitrarily close to the right-hand side of (13) for small enough $|x|$. Hence, we see that we cannot write a smaller value in place of $1/\sqrt{5}$ on the right-hand side of (13). This proves Theorem 16. We also showed that $\alpha = \left(\sqrt{5}-1\right)/2$ is an example for which the bound in Theorem 16 cannot be improved.[19]

We have shown that in a lattice of the form (12), $|xy|$ cannot be less than $d/\sqrt{5}$ for any lattice point other than the origin. Therefore, applying

---

[19] Generalizing this procedure to an arbitrary lattice we get the following: Let m_x denote the lower bound of the product $|xy|$, for those lattice points for which $|x|$ is smaller than some bound, and m_y the lower bound of the product for those lattice points for which $|y|$ is bounded. Then, assuming that a, respectively b', is not 0, m_x is a lower bound for the lattice

$$x = u + \frac{b}{a}v, \qquad y = dv$$

when $|x|$ is bounded, while m_y is a lower bound for the lattice

$$x = du, \qquad y = \frac{a'}{b'}u + v$$

when $|y|$ is bounded (d is still the area of an empty parallelogram).

This means that with fixed d, m_x depends only on the linear form x, and m_y depends only on the linear form y. This is also true in the case where there are infinitely many lattice points on one of the axes, since assuming that this is the case for the x-axis, then the ratio of the coefficients of the linear form y are rational. Then $m_y = 0$, and this is again determined only by the linear form y. (This is also the case when $b' = 0$, which was assumed above not to happen, similarly when $a = 0$, in which case $m_x = 0$.)

the proof of Theorem 16 to the x-axis instead of the y- axis, we still get a minimum of $d/\sqrt{5}$. Reviewing the proof, it follows that $\left(\sqrt{5}+1\right)/2$ is also an example for which the bound in Theorem 16 cannot be improved.

The theorem is due to HURWITZ. The approximation of two or more real numbers $\alpha_1, \alpha_2, \ldots, \alpha_n$ by fractions with a common denominator $u_1/v, u_2/v, \ldots, u_n/v$ has been investigated too. It can be proved that there exists a positive constant $c_n < 1$ and infinitely many fractions for which

$$\left|\alpha_i - \frac{u_i}{v}\right| \leq \frac{c_n}{v^{1+1/n}} \quad (i = 1, 2, \ldots, n).$$

However, an arbitrarily small c_n cannot be given such that the corresponding approximation is good for every $\alpha_1, \alpha_2, \ldots, \alpha_n$. By Theorem 16, the lowest bound for c_1 is $1/\sqrt{5}$, but the lowest bound for any c_n for $n > 1$ is still unknown.

32. With a slight change in the proof of Theorem 16, we get another interesting result. We modify slightly the notation. Rename the lattice point P_2 of Figure 23 to Q and denote by P_2 the lattice point on the lattice line P_0Q in the first quadrant closest to the y-axis, and let its coordinates be x_2 and y_2. It is possible that P_2 coincides with Q. We have $\boldsymbol{P_0P_2} = k_1\boldsymbol{OP_1}$ (where bold indicates vectors rather than lines) for some positive integer k_1 (Figure 24). Then

$$x_2 = x_0 - k_1x_1, \qquad y_2 = y_0 + k_1y_1,$$

$$M_2 = x_2y_2 = M_0 - k_1^2M_1 + k_1(x_0y_1 - x_1y_0),$$

and (9) is replaced by

$$-M_0 + k_1^2M_1 + M_2 = k_1(x_0y_1 - x_1y_0).$$

Equation (10) is still valid, and hence

$$k_1d = k_1(x_0y_1 + x_1y_0).$$

Taking the difference of the squares of corresponding sides we have the equality

$$k_1^2d^2 = k_1^2M_2(2M_0 + k_1^2M_1 + 2M_2) + (M_2 - M_0)^2.$$

Denoting the smallest of M_0, M_1, and M_2 by M_1', we omit the second term on the right-hand side and simplify by k_1^2 so we get

$$d^2 \geq (k_1^2 + 4)M_1'^2 \qquad \text{i.e.,} \qquad M_1' \leq \frac{d}{\sqrt{k_1^2 + 4}}.$$

Since $k_1 \geq 1$, we essentially get Theorem 16 once more, but also something more.

We may continue this procedure such that we choose P_3 to be the lattice point closest to the y-axis in the second quadrant on the lattice line going

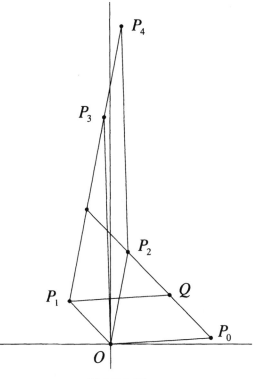

FIGURE **24.**

through P_1, parallel to OP_2. Next pick P_4 to be the lattice point closest to the y-axis in the first quadrant on the lattice line going through P_2, parallel to OP_3, and so on. The intervals P_1P_3, P_2P_4, and the following ones get divided into $k_2, k_3, \ldots$ parts by the lattice points on them, where each k_i is a positive integer $(i = 2, 3, \ldots)$. Denote the absolute value of the product of coordinates of P_i by M_i, and the smallest of M_{i-1}, M_i, and M_{i+1} by M_i'.

Repeating the above procedure for the sequence of triangles

$$P_1P_2P_3, P_2P_3P_4, \ldots$$

we get that

$$M_i' \leq \frac{d}{\sqrt{k_i^2 + 4}} \quad (i = 1, 2, \ldots).$$

Since all the k_i's are at least 1 and at least every third M_i' belongs to a different lattice point, we have another proof of both claims of Theorem 16. In addition, we have also developed a procedure for finding infinitely many appropriate lattice points.

The true gain, however, is that if one of the k_i's is not 1 for some lattice, then for the corresponding M_i' we have

$$M'_i \leq \frac{d}{\sqrt{8}}.$$

If infinitely many k_i's are greater than 1, then the above relation is also true for infinitely many lattice points. Since in a lattice of type (12) this is not satisfied for any of the lattice points (other than the origin), we see that for this lattice, $k_i = 1$ for all i. However, since we have seen that only for those lattices we cannot remove the equality in Theorem 16, for the other lattices we have found lattice points satisfying the strict inequality. Summarizing, we have the following theorem.

Theorem 16′ *In lattices not of type* (12), *there exist lattice points whose coordinates satisfy*

$$|xy| \leq \frac{d}{\sqrt{8}}.$$

It was A. A. MARKOV who discovered the phenomenon that not considering lattices for which equality holds in Theorem 16, a sharper inequality holds for the rest of the lattices; i.e., there is no lattice for which the lower bound of the absolute value of the product of coordinates of its lattice points would be a number between $d/\sqrt{5}$ and $d/\sqrt{8}$.

We can determine again the lattices for which Theorem 16′ cannot be improved, and for the others yet again a larger number can be written in the denominator. These decreasing minima and the corresponding lattices form an infinite sequence, where the minima tend to $d/3$.[20]

33. In the previous sections the origin was always a lattice point. These lattices are called *homogeneous*. Some problems lead, however, to planar lattices without this restriction. In this case we call the lattice *inhomogeneous*.

Usually, for inhomogeneous lattices it is supposed that the axes do not contain any lattice points. This restriction can be omitted, and homogeneous lattices can be considered as special cases of inhomogeneous ones.

In some cases, however, as in the next theorem, it is necessary to make an assignment of each axis to one of the quadrants, so that every point in the plane belongs to only one quadrant. For instance, let the positive part of the x-axis belong to the fourth quadrant ($x > 0, y \leq 0$), the positive part of the y-axis to the second quadrant ($x \leq 0, y > 0$), and the negative parts of both axes along with the origin to the third quadrant ($x \leq 0, y \leq 0$). The third quadrant is then closed, while the first ($x > 0, y > 0$) is open. With such a convention, quadrants are convex and do not have common elements.

If the coordinates of a lattice point are real numbers k and l, then the coordinates of an arbitrary lattice point can be written as

---

[20] See A. A. MARKOFF: *Math. Annalen* **15** (1879), pp. 381–406. For a different proof, further references, and problems, see the book of J. W. S. CASSELS referred to in footnote 4, pp. 18–44.

$$x = au + bv + k, \qquad y = a'u + b'v + l,$$

where u and v are integers, and $d = |ab' - a'b| > 0$ is the area of the empty lattice parallelograms (k and ℓ are fixed).

An empty parallelogram whose vertices belong to distinct quadrants is called a *divided cell*. Note that a divided cell always contains the origin. In the homogeneous lattice, according to our conventions, every divided cell's vertex belonging to the third quadrant is the origin.

In numerous problems regarding inhomogeneous lattices, the following theorem, due to DELONE (sometimes written DELAUNAY), proves to be a useful tool.

Theorem 17 *All inhomogeneous lattices contain a divided cell.*

For the proof, let us pick two lattice points A and B in the first and second quadrants. Then drop perpendiculars from these to the x-axis. The trapezoid we get this way contains at most finitely many lattice points above the x-axis. From these we can pick two such that the new trapezoid does not contain additional lattice points from either the first or the second quadrant. Without loss of generality, we may assume that A and B are already such lattice points. Then A and B are neighbors on the lattice line through them (Figure 25).

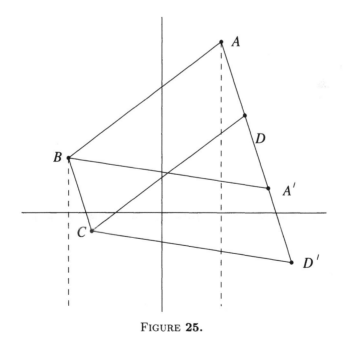

FIGURE **25.**

Extend the perpendiculars from these points beyond the x-axis. The strip between the two extended perpendiculars contains segments of lattice lines parallel to AB of length AB; hence it has lattice points (in the interior or on the endpoints of the segments). On the first such line, let C and D be neighboring lattice points intercepting the y-axis.

One of C and D is within the strip and below the x-axis. If the other one is also below the x-axis, then $ABCD$ is a divided cell. If, for instance, D is above the x-axis, then it is not inside the strip; i.e., it lies further from the y-axis than A. Hence the lattice line AD intersects the x-axis on its positive side. Denote the two neighboring lattice points on this line intercepting the x-axis by A' and D'. Then A', B, C, and D' all lie in different quadrants.

The parallelogram $ABCD$ is empty, since the lattice lines AB and CD are neighboring, and hence AD and BC are also neighboring. So the lattice parallelogram $A'BCD'$ is also empty, therefore a divided cell, by our previous observation.

If it is not D but C above the x-axis, then proceeding as above using BC instead of AD, we find a divided cell similarly as before. This proves Delone's theorem.

34. The history of the previous theorem is rather interesting as well. B. N. DE-LONE himself gave a discussion of the problem, with many interesting details.[21] D. B. SAWYER cites the theorem as something obvious, unaware of DELONE's work. Thereafter, many proofs appeared about the existence of divided cells, still without reference to DELONE's work. Finally, BARNES and SWINNERTON-DYER found the article and developed some applications. Later, BIRCH published an example of a lattice in three dimensions in which there exists no divided cell, again not knowing that this counterexample, too, appears in DELONE's article.[22]

It would be of interest to determine all parallelepiped lattices that do not contain a divided cell.

35. Starting from a divided cell, we can get a good overview of all divided cells of the lattice in the following way. If the sides of the cell intersecting the x-axis are not parallel to the y-axis, then take the lattice points on the lattice lines through them intercepting the y-axis. These are neighboring lattice points on neighboring lattice lines belonging to different quadrants, i.e., vertices of a divided cell. This procedure can be repeated infinitely unless we get a cell for which two of its sides are parallel to the y-axis.

The procedure can be repeated in a similar fashion in the opposite direction as well, if the sides intersecting the y-axis are not parallel to the x-axis.

---

[21] B. N. DELONE: *Izvesztyija Akad. Nauk SSSR. Ser. Mat.* **11** (1947), pp. 505–538 (in Russian). In German, see *Sowjetwissenschaft* **2** (1948), pp. 178–210. Cf. also J. SURÁNYI: *Acta Sci. Math. Szeged* **22** (1961), pp. 85–90.

[22] D. B. SAWYER: *Journal London Math. Soc.* **23** (1948), pp. 250–251; E. S. BARNES, H. P. F. SWINNERTON-DYER: *Acta Mathematica* **87** (1952), pp. 259–323; *ibid.* **88** (1952), pp. 279–316; and *ibid.* **92** (1954), pp. 199–234. B. J. BIRCH: *Proc. Cambridge Phil. Soc.* **53** (1956), p. 536.

In the special case where there exists an empty lattice parallelogram with sides parallel to both axes, then only one divided cell exists.

This way we have obtained a sequence of divided cells that is infinite in the directions of both axes, or infinite in the direction of only one axis, or finite in both directions. DELONE also showed that this sequence contains all divided cells (see Exercise 28).

36. MINKOWSKI determined the minimum of the product $|xy|$ for lattice points of inhomogeneous lattices.

Theorem 18 *Any inhomogeneous lattice in which the area of the empty lattice parallelograms is d has a lattice point with coordinates (x, y) for which*

$$|xy| \le \frac{d}{4}.$$

It is reasonable to look for such a lattice point among the vertices of a divided cell $ABCD$. Let E be the center of the cell. The diagonals of the parallelogram divide it into four triangles of equal area. At least one of these includes the origin. Assume that this is the triangle EAB (Figure 26). Then the area of triangle OAB is not greater than $d/4$.

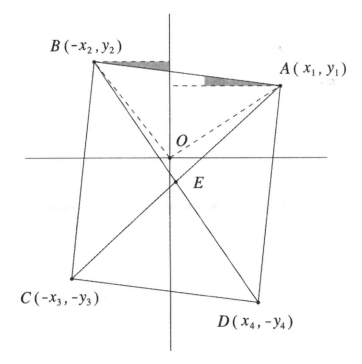

FIGURE **26.**

Drawing perpendiculars from A and B onto the half-axis crossing the triangle produces with point O two right triangles. Triangle EAB contains at least one of these right triangles, and the other one extends beyond EAB, unless, of course, AB is perpendicular to that axis, in which case the two right triangles cover OAB. In the latter case, the area of one of the triangles is not greater than $d/8$, and for the corresponding lattice point the inequality in Theorem 18 is satisfied. Equality holds only when O and E coincide and the axis is the perpendicular bisector of AB.

If AB is not perpendicular to the axis that intersects it, but the part of the triangle OAB containing the right triangle—say the part corresponding to vertex A—does not have larger area than the other part, then the area of this right triangle is less than $d/8$, and vertex A satisfies the claim of Theorem 18.

However, if the part containing the right triangle has greater area than the other part, then it also contains the midpoint of the line segment AB. We see that the mirror image of the part of the right triangle extending beyond the cell with respect to the midpoint of AB is in the triangle OAB. Thus we see that the area of both right triangles does not fill OAB. Hence the area of one of them is at most $d/8$. For the corresponding lattice point, the claim of the theorem holds.

Following the proof more closely, we can see that equality holds only when O is the center of the cell, and one of the axes perpendicularly bisects one of its sides. It is obvious that in this case, the other axis is also a perpendicular bisector of the other side, and hence the divided cell is a rectangle with sides parallel to the axes whose midpoint is the origin. Thus in addition to Theorem 18, we have shown that one cannot write a smaller value in place of $d/4$; specifically, we have shown when equality holds.

37. In the following we present SAWYER's elegant proof of Theorem 18 referred to in Section 34. If a lattice point were to fall on one of the axes, then the product of the coordinates of such a lattice point would be 0, and the claim of Theorem 18 is true.

If this is not the case, then denote the coordinates of the vertices of the divided cell as in Figure 26. Furthermore, let $m_i = x_i y_i$ $(i = 1, 2, 3, 4)$. We calculate the area of the parallelogram as the sum of the areas of the triangles OAB, OBC, OCD, and ODA. In turn, the areas of these triangles can be expressed using analytic geometry. We have the following:

$$d = \frac{1}{2}(x_1 y_2 + x_2 y_1 + x_2 y_3 + x_3 y_2 + x_3 y_4 + x_4 y_3 + x_4 y_1 + x_1 y_4)$$

$$= \frac{1}{2}\left(m_1\frac{y_2}{y_1} + m_2\frac{y_1}{y_2} + m_2\frac{y_3}{y_2} + m_3\frac{y_2}{y_3} + m_3\frac{y_4}{y_3} + m_4\frac{y_3}{y_4} + m_4\frac{y_1}{y_4} + m_1\frac{y_4}{y_1}\right).$$

Replacing all the m_i's by the smallest of them, denoted by m, makes the sum smaller, or at least not greater. Also, using the inequality $s + 1/s \geq 2$, valid for all positive numbers s, we get that $d \geq 4m$. This means that the claim of Theorem 18 holds true for at least one of the vertices of the divided cell.

Following the proof more closely, one can also determine when equality holds. We leave this to the reader (Exercise 31).

Exercises:

28. Prove that the procedure of getting new divided cells from a given one gives rise to all the divided cells.

29. Prove that the second claim of Theorem 16 is true for $\left(\sqrt{5} + 1\right)/2$ as well.

30. Determine those lattices for which the estimate of Theorem 16' cannot be improved.

31. Following SAWYER's proof, determine the cases where the equality cannot be dropped in Theorem 18.

5. Properties of Prime Numbers

1. We have seen two proofs in Chapter 2 showing the uniqueness of prime decompositions of integers (Theorem 1.8). The consequences of that theorem there and also in following chapters make it clear why Theorem 1.8 is called the fundamental theorem. The prime numbers mentioned in that theorem are distributed among the integers in a very peculiar way. One can get a feel for this by looking at the sequence of primes less than 150. We list these primes, writing the differences between consecutive primes below them, and writing those differences that are larger than all the previous differences in boldface.

$$
\begin{array}{ccccccccccccc}
2 & 3 & 5 & 7 & 11 & 13 & 17 & 19 & 23 & 29 & 31 & 37 & 41 & 43 \\
& \mathbf{1} & 2 & 2 & \mathbf{4} & 2 & \mathbf{4} & 2 & 4 & \mathbf{6} & 2 & 6 & 4 & 2 & 4
\end{array}
$$

$$
\begin{array}{cccccccccccc}
47 & 53 & 59 & 61 & 67 & 71 & 73 & 79 & 83 & 89 & 97 & 101 \\
& 6 & 6 & 2 & 6 & 4 & 2 & 6 & 4 & 6 & \mathbf{8} & 4 & 2
\end{array}
$$

$$
\begin{array}{cccccccccc}
103 & 107 & 109 & 113 & 127 & 131 & 137 & 139 & 149 \\
& 4 & 2 & 4 & \mathbf{14} & 4 & 6 & 2 & 10
\end{array}
$$

It is not surprising that since the earliest times both professional and amateur mathematicians have shown great interest in properties of prime numbers. We have seen EUCLID's proof (among others) that there are infinitely many prime numbers (Theorem 1.4 and Exercise 1.19), that the difference between two consecutive primes can be arbitrarily large (Exercise 1.20), and that there are so-called *reclusive* primes, primes that are arbitrarily far from all other primes (Theorem 2.8 and Exercise 2.18).

Outsiders pose such questions as, "Has the secret of prime numbers been discovered?" But it is not clear what type of answer they hope for. If they are hoping for a formula that generates prime numbers, then the answer is yes. MILLS[1] proved that there exists a number c, for which $[c^{3^n}]$ is a prime number for every positive integer n, SIERPIŃSKI[2] presented a formula which gives the nth prime number for every n.

---

[1] W. H. MILLS: *Bulletin Amer. Math. Soc.* **53** (1947), p. 604.
[2] W. SIERPIŃSKI: *Comptes Rendus Acad. Sci.*, Paris **235** (1952), pp. 1078–1079.

These are very interesting results, but they are not useful in practice. MILLS's formula uses the prime numbers to determine the constant c; moreover, it applies the deep theorem that between the cubes of two consecutive integers (beyond a certain point) there is always a prime number. SIERPIŃSKI's formula is similar. The results we proved earlier, which are less precise, and the above-mentioned theorem tell us more about the primes than these two formulas. In the remainder of this chapter we will be dealing with similar types of theorems, many of which can be proved by elementary methods.

2. Unless we specify otherwise, we will denote by p_n the nth prime number, ordered by size ($p_1 = 2, p_2 = 3, p_3 = 5$, etc.), and by d_n the difference between consecutive primes: $d_n = p_{n+1} - p_n$. We have seen that the second sequence is not bounded; in fact, there are arbitrarily large consecutive terms. The value 1 occurs only once, since all even numbers greater than 2 are composite. The difference 2 occurs often between consecutive primes. These primes are consecutive odd numbers and are called *twin primes*. Looking at any list of prime numbers, one sees that twin primes occur often, even among larger prime numbers. KRAITCHIK determined the prime numbers occurring among the 10 000 numbers preceding and following one billion, and even here twin primes occur. It is, however, unknown whether there are infinitely many twin primes.

It is easy to show (see Exercise 1) that for infinitely many n, $d_{n+1} > d_n$. On the other hand, it is not at all easy to prove that for infinitely many $n, d_{n+1} < d_n$.[3]

It is expected that the inequalities $d_{n+1} > d_n$ and $d_{n+1} < d_n$ both occur roughly half of the time up to any bound. It is also possible that equality $d_{n+1} = d_n$ occurs infinitely often. It has been proved only that there exist positive numbers h and c such that for all n less than a sufficiently large bound N, the inequality $d_{n+1} > (1 + h)d_n$ holds for cN values of n.[4]

It is still unknown whether either of the inequalities $d_{n+2} > d_{n+1} > d_n$ or $d_{n+2} < d_{n+1} < d_n$ occurs infinitely often. (If both were to occur only finitely many times, this would mean that after a certain point, the sequence $d_{n+1} - d_n$ would alternate in sign, but this most certainly does not happen.)

More recently, Maier proved that the difference $d_{n+1} - d_n$ takes on values less than $0.2484 \log d_n$ infinitely often.[5]

It is unknown what lower bounds can be given for d_n. We have mentioned that it is unknown whether the difference 2 occurs infinitely often (twin primes) among the d_n's; it is also unknown whether arbitrarily small numbers occur among the quotients $d_n / \log n$. All that has been proved in

---

[3] P. ERDŐS and P. TURÁN: *Bulletin Amer. Math. Soc.* **54** (1948), pp. 371–378. They were the first to consider questions relating to d_n and the first to present results.

[4] P. ERDŐS: *Publicationes Math. Debrecen* **1** (1949–1950), pp. 33–37.

[5] H. C. MAIER: *Michigan Math. Journ.* **35** (1988), pp. 323–344.

this direction is that there is a constant c such that the quotient is less than $1 - c$ for infinitely many n.[6]

We have seen that there are reclusive primes, which means that two consecutive d_n's can both be larger than any given bound. An even stronger result is known: The quotient

$$\frac{\min(d_n, d_{n+1})}{\log n}$$

takes on arbitrarily large values.[7] Results for the minimum of more than two consecutive terms were not found until more than 30 years later. MAIER showed that for an arbitrary k, there exists a constant c for which there are infinitely many n's such that the quotient

$$\frac{\min(d_{n+1}, d_{n+2}, \ldots, d_{n+k})(\log \log \log n)^2}{\log n \log \log n \log \log \log \log n}$$

is greater than c; moreover, k can increase as n increases.[8]

It is an old conjecture that for n large enough, d_n is smaller than n to an arbitrarily small positive power; in fact, it is probably smaller than a constant times $(\log n)^2$. In this direction, the known "record" is by R. C. BAKER and HARMAN,[9] who proved the bound $n^{0.535}$.

In the other direction we know only that there is a constant c for which infinitely many n's satisfy the inequality[10]

$$d_n > c\, \frac{\log n \log \log n \log \log \log \log n}{(\log \log \log n)^2}.$$

The conjectures concerning upper bounds on the d_n's would imply that for any positive δ, there is a prime number between $n^{1+\delta}$ and $(n+1)^{1+\delta}$ if n is large enough. The bound mentioned above gives the best result of this type, which shows that between any two large enough consecutive cubes there is a prime.[11] The equivalent statement for squares looks quite difficult.

Exercises:

[6] R. A. RANKIN: *Journ. London Math. Soc.* **22** (1947), pp. 226–230.

[7] P. Erdős: *Bull. Amer. Math. Soc.* **54** (1948), pp. 885–889.

[8] H. MAIER: *Advances in Math.* **39** (1981) pp. 257–269.

[9] R. C. BAKER, G. HARMAN: The difference between consecutive primes, *Proc. London Math. Soc.* (3) **72**, 1996, pp. 261–280.

[10] P. ERDŐS: *Quarterly Journ. of Math., Oxford Ser.* **6** (1935), pp. 124–128. To date the largest communicated value of c is $C \cdot e^\gamma$, where γ is the Euler–Mascheroni constant and $C \approx 1.31$. H. MAIER and C. POMERANCE: *Transactions of the Amer. Math. Soc.*, **322** (1990), pp. 201–237. J. PINTZ showed that C can be improved to a constant larger than 2. Verbal communication.

[11] R. A. RANKIN: *Journ. of the London Math. Soc.* **13** (1938), pp. 242–247. The result of footnote 8 is somewhat stronger than this.

1. Prove that for infinitely many $n, d_{n+1} > d_n$.

2*. Prove that there are infinitely many twin primes if and only if there are infinitely many integers that cannot be written in any of the following forms:

$$6uv + u + v, \quad 6uv + u - v, \quad 6uv - u + v, \quad 6uv - u - v,$$

for positive integers u and v (S. Golomb).

3. We start our further investigations with simple questions. In EUCLID's proof that there are infinitely many primes, we added 1 to the product of the given primes to give a number divisible by new primes. Repeating this procedure, we can produce arbitrarily many primes. In other ways, we can also produce sequences of pairwise relatively prime numbers greater than 1, and then choosing a prime divisor of each, we again produce infinitely many primes. The following theorem, also based on EUCLID's ideas, produces such a sequence.

Theorem 1 *If b and c are arbitrary natural numbers that are relatively prime, then the elements of the sequence defined by the recursion*

$$a_0 = b, \quad a_{n+1} = a_0 a_1 \cdots a_n + c \quad (n = 0, 1, \dots)$$

are pairwise relatively prime.

We prove the theorem indirectly. Suppose that the sequence contains elements that are not relatively prime. Let a_i be the first element that is not relatively prime to some other element a_j of the sequence. These two elements have a prime common divisor, which we will denote by p. One of the terms of the product on the right-hand side of

$$a_j = a_0 a_1 \cdots a_{j-1} + c$$

is a_i itself, and hence the product is divisible by p as well. On the other hand, p divides a_j, and therefore it must divide c, too.

Thus the left-hand side of

$$a_i = a_0 a_1 \cdots a_{i-1} + c$$

and the second term of the right-hand side are both divisible by p, therefore the product on the right-hand side is as well. But according to the prime property, this is possible only if a_k is also divisible by p, for some k less than i. Hence, a_k would have a common divisor (greater than one) with another element of the sequence, but this contradicts our assumption that a_i is the first such element. Therefore, Theorem 1 must hold true.

4. For $b = 3$ and $c = 2$, we can see by induction that we get the sequence

$$a_n = 2^{2^n} + 1 \qquad (n = 0, 1, \dots),$$

hence this sequence is made up of pairwise relatively prime integers. This can also be seen directly. These numbers have at least one prime divisor each, showing that p_n is not larger than a_{n-1}.

From these results we can also establish a lower bound on the number of primes not larger than x. We will call this quantity $\pi(x)$. It is easy to see that this result implies that $\pi(x)$ is not smaller than $\log \log x$ times a positive constant. We can immediately give a better bound for this, and later we will prove a significantly larger lower bound.

We can also produce other such sequences whose elements are pairwise relatively prime, but no such sequence is known that would give a significantly better lower bound. It would be interesting to decide whether there is some mathematical reason for this. Denser sequences of relatively prime numbers of course exist, for instance the sequence of prime numbers itself. In fact, this is the densest such sequence. Namely, if the sequence $a_1 < a_2 < \cdots$ is made up of pairwise relatively prime elements larger than 1, then each element must contain a new prime, and a_n must be at least as large as the nth prime.

Exercises:

3. (a) Prove that if $b = 3$ and $c = 2$ for the sequence in Theorem 1, then $a_n = 2^{2^n} + 1 \ (n = 0, 1, \dots)$.

(b) Prove without using Theorem 1 that the elements of the sequence above are pairwise relatively prime.

4. Prove that if c and d are relatively prime positive integers, then the elements of the sequences generated by the following recursions are pairwise relatively prime:

(a) $a_0 = c, \quad a_1 = d, \quad a_{n+1} = \prod_{1 \leq 2k-1 \leq n} a_{2k-1} + \prod_{0 \leq 2k \leq n} a_{2k}$;

(b) $a_0 = c, \quad a_1 = c + d, \quad a_{n+1} = a_n(a_n - d) + d \quad (n = 1, 2, \dots)$.

5. In the proof that there are infinitely many primes given in Exercise 1.19 (b), the basic idea is that every number can be written as a product of a square and a square-free number (a number not divisible by a square larger than 1). Estimating the numbers of the two types of factors and assuming that there are only finitely many prime numbers, we can see that we cannot produce enough numbers in this way up to a large enough bound.

If we omit the indirect hypothesis, this idea can be used to give a lower bound on $\pi(x)$, the number of primes not greater than x. We can prove the following result:

Theorem 2 *There exists a positive constant c for which the inequality*

$$\pi(x) > c \log x$$

holds for all $x > 2$.

Applying the above idea, let $M(x)$ be the number of squares up to x. We have

$$M(x) = \left[\sqrt{x}\right] \leq \sqrt{x}.$$

Let $N(x)$ be the number of square-free integers up to x. The canonical decompositions of these are products of primes not larger than x, each prime to the 0th or 1st power. The number of these primes is $\pi(x)$, so these give $2^{\pi(x)}$ square-free numbers.[12] Therefore,

$$N(x) \leq 2^{\pi(x)}.$$

From these we get the following inequality concerning the number of integers not larger than x:

$$[x] \leq M(x)N(x) \leq \sqrt{x}2^{\pi(x)},$$

from which for $x > 2$, it follows that

$$2^{\pi(x)} \geq [x]/\sqrt{x} > (x-1)/\sqrt{x} > \frac{1}{2}\sqrt{x},$$

and hence

$$\pi(x) > \frac{1}{2}\log x - \log 2 > c\log x,$$

giving a c slightly less than $\frac{1}{2}$, satisfying the claim of the theorem.

6. We can measure how dense a sequence of natural numbers is among the natural numbers by considering the reciprocals of the terms of the original sequence and determining whether the sum of the first n terms of this is always less than some bound, or whether these initial sums grow beyond all bounds. In the first case, the reciprocals of the terms decrease rapidly hence the terms grow quickly. In the second case the reciprocals decrease slowly, hence the sequence does not increase too quickly.

Both of the two cases occur. Considering all the natural numbers, we have the series

$$1 + \frac{1}{2} + \frac{1}{3} + \cdots + \frac{1}{n},$$

known as the harmonic series, which diverges by Fact 6. We may conversely consider the sequence of squares, which gives rise to the series

---

[12] Most of these are greater than x. From the 8 primes not greater than 20, for example, one can form 256 square-free numbers, but of these only 13 are not greater than 20.

$$1 + \frac{1}{2^2} + \frac{1}{3^2} + \cdots + \frac{1}{n^2} < 1 + \frac{1}{1 \cdot 2} + \frac{1}{2 \cdot 3} + \cdots + \frac{1}{(n-1)n}$$

$$= 1 + \left(1 - \frac{1}{2}\right) + \left(\frac{1}{2} - \frac{1}{3}\right) + \cdots + \left(\frac{1}{n-1} - \frac{1}{n}\right)$$

$$= 2 - \frac{1}{n} < 2.$$

This property holds, of course, for any subsequence of squares, such as the sequence of squares of primes; taking any number of elements of this sequence, we see that the sum of its reciprocals is less than 2. (In Exercise 5* below, we will need a better bound for the sum of reciprocals of squares of primes.)

7. EULER showed that the sequence of prime numbers is a dense sequence in the sense defined above. We will present two proofs of this theorem. The first is a short indirect proof due to ERDŐS.[13]

Theorem 3 *The sum of the reciprocals of the prime numbers diverges.*

Suppose the contrary. Then we can choose a bound K such that the sum of the reciprocals of primes larger than K is less than $\frac{1}{2}$. Let r be the number of primes less than or equal to K.

Now let N be an arbitrarily large integer greater than K. We shall reach the absurd conclusion that N must, in fact, be bounded.

We divide the set of integers less than N into two groups. The first group, containing, say, N_1 integers, will consist of the integers less than N that are divisible by a prime greater than K. Then for each prime between K and N there are at most $[N/p]$ such multiples, and so we obtain the estimate

$$N_1 \leq \sum_{K < p \leq N} \left[\frac{N}{p}\right] \leq N \cdot \sum_{K < p \leq N} \frac{1}{p} < \frac{1}{2}N,$$

where the last inequality follows from our assumption that $\sum_{K < p \leq N} \frac{1}{p}$ is less than $\frac{1}{2}$.

The second group contains the remaining, say, N_2 integers less than N that are not divisible by a prime greater than K. We write each such integer as the product of a square-free term and a square, following the ideas in Section 5. There are at most $\left[\sqrt{N}\right]$ possible values for the square part of the product. The square-free part has a canonical decomposition consisting of the first r primes, each raised to either the zeroth or the first power; hence there are at most 2^r of these.[14] Therefore, $N_2 < \left[\sqrt{N}\right] \cdot 2^r$.

---

[13] P. ERDŐS: *Oberdruk uit Math. B.* **7** (1938), pp. 1–2. One can find another proof here as well; see Exercise 5 of this reference.

[14] In making this estimate we do not pay attention to whether or not the first term has a prime factor greater than K, and whether or not the second term is greater than N.

Combining the estimates for N_1 and N_2, we get the following inequality:

$$N_1 + N_2 = N < \frac{1}{2}N + \left[\sqrt{N}\right]2^r \le \frac{1}{2}N + \sqrt{N}2^r,$$

and $N < \frac{1}{2}N + \sqrt{N}2^r$ immediately yields

$$\frac{1}{2}\sqrt{N} < 2^r.$$

This leads to an immediate contradiction, since the right-hand side is a fixed value for a given K, while N can be arbitrarily large. Thus Theorem 3 is true.

Exercises:

5*. (a)   Show that the sum of the reciprocals of the squares of primes is bounded by a value less than $\frac{1}{2}$.

(b*)   Modify the proof of Theorem 3 such that we estimate the number of integers that are not divisible by primes greater than K separately when they are divisible by a square greater than one, and separately those that are square-free.

6.   Prove that the sequence of natural numbers all of whose prime divisors are greater than some natural number K is made up of finitely many residue classes modulo the product of primes less than K.

7.   We are given a sequence of distinct natural numbers, and we know that for an appropriate positive number c, there exist arbitrarily large numbers x such that the number of terms not greater than x is at least cx. Prove that the sum of the reciprocals of enough elements can be arbitrarily large.

8. The first proof of Theorem 3 is due to EULER, who also gave a lower bound for the sum in the theorem. He noticed that this sum is closely related to the logarithm of the following product:

$$S_k = \prod_{i=1}^{k} \frac{1}{1 - \frac{1}{p_i}},$$

where p_k is the largest prime not greater than x. Applying the inequality in Fact 5 in two different ways to the logarithm of the reciprocals of the terms in the denominator of the above quotient (first applying it to an arbitrary number p greater than 1), we have that

$$-\frac{1}{p} \ge \log\left(1 - \frac{1}{p}\right) = -\log\left(1 - \frac{1}{p}\right)^{-1} = -\log\left(1 + \frac{1}{p-1}\right) \ge -\frac{1}{p-1}.$$

Multiplying this by -1 and subtracting $1/p$, we get that

$$0 \le \log\left(1 - \frac{1}{p}\right)^{-1} - \frac{1}{p} \le \frac{1}{p(p-1)}.$$

Now replace p by $p_1, p_2, \ldots, p_k$, and sum up the terms. The sum of the logarithms is $\log S_k$, and hence it follows that

$$0 \le \log S_k - \sum_{i=1}^{k} \frac{1}{p_i} \le \sum_{i=1}^{k} \frac{1}{p_i(p_i - 1)} \le \sum_{n=2}^{[x]} \frac{1}{n(n-1)}$$

$$= \sum_{n=2}^{[x]} \left(\frac{1}{n-1} - \frac{1}{n}\right) = 1 - \frac{1}{[x]} < 1.$$

Based on this, the investigation of the sum can indeed be substituted by the investigation of S_k.

9*. We decrease the factors of the product S_k so that they can be written as a geometric series. For an arbitrary real number p different from 1 and a positive integer r,

$$\frac{1}{1 - \frac{1}{p}} > \frac{1 - (\frac{1}{p})^{r+1}}{1 - \frac{1}{p}} = 1 + \frac{1}{p} + \frac{1}{p^2} + \cdots + \frac{1}{p^r}.$$

We replace the p by p_i and r by r_i for all p_i and r_i satisfying

$$p_i^{r_i} \le x < p_i^{r_i+1}.$$

In this way we get

$$S_k > \prod_{i=1}^{k} \left(1 + \frac{1}{p_i} + \frac{1}{p_i^2} + \cdots + \frac{1}{p_i^{r_i}}\right).$$

The product on the right-hand side consists of terms whose reciprocals are products of the form

$$p_1^{s_1} p_2^{s_2} \cdots p_k^{s_k},$$

where

$$0 \le s_i \le r_i \quad (i = 1, 2, \ldots, k).$$

Among these, every integer not greater than x occurs, and hence

$$S_k > 1 + \frac{1}{2} + \frac{1}{3} + \cdots + \frac{1}{[x]}.$$

The right-hand side is a partial sum of the harmonic series, and by Fact 6 it is larger than $\log([x] + 1)$. Summarizing our results, we have

$$1 + \sum_{p \leq k} \frac{1}{p} = 1 + \sum_{i=1}^{k} \frac{1}{p_i} > \log S_k > \log \log([x] + 1) > \log \log x.$$

In this way we have the following result:

Theorem 4 *The sum of the reciprocals of all primes not larger than x satisfies the inequality*

$$\sum_{p \leq x} \frac{1}{p} > \log \log x - 1.$$

By similar methods, using for example the canonical decomposition of $n!$ given in Chapter 1, one can give an upper bound to the given sum, and the two bounds together give that

$$\left| \sum_{p \leq x} \frac{1}{p} - \log \log x - 1 \right|$$

is bounded by a constant independent of x. Further, using elementary methods it is possible to produce good bounds for expressions that imply useful results for the distribution of prime numbers. We will, however, not continue with these. Instead, we will turn to significantly improved bounds for $\pi(x)$, as promised earlier.

10. We have seen several examples of how irregularly and peculiarly primes follow each other. The previous section showed that even with all these peculiarities we can still determine certain regularity. The number $\pi(x)$ of primes not larger than x was studied already in the eighteenth century, when it was noted that for large x, the number grows approximately like $x/\log x$. More precisely, the quotient

$$\frac{\pi(x)}{x/\log x}$$

is arbitrarily close to 1 for x large enough.

The first proof of anything in this direction was given by CHEBYSHEV in the middle of the nineteenth century, who showed that there exist positive constants c_1 and c_2 for which

$$\frac{c_1 x}{\log x} < \pi(x) < \frac{c_2 x}{\log x},$$

hence that the previously mentioned quotient falls between positive bounds. In the following we will prove these inequalities.

The lower bound can be obtained fairly easily based on certain properties of the binomial coefficient $\binom{n}{k}$, specifically that no prime power larger than n occurs in its canonical decomposition (Theorem 1.13). Applying this to the binomial coefficient $\binom{2n}{n}$, we get

$$\binom{2n}{n} = \prod_{p \leq 2n} p^{\beta} \leq (2n)^{\pi(2n)}.$$

We next estimate this binomial coefficient from below. We start by cancelling an $n!$ from the denominator with all the even terms in the numerator, leaving a factor of 2^n. We then replace all odd terms greater than 1 in the numerator by the even term 1 less than it. This gives

$$\binom{2n}{n} = \frac{(2n)!}{(n!)^2} = \frac{2 \cdot 3 \cdots (2n-1)2n}{1 \cdot 2 \cdots n \cdot n!} = 2^n \cdot \frac{3 \cdot 5 \cdots (2n-1)}{n!}$$

$$> 2^n \cdot \frac{2 \cdot 4 \cdots (2n-2)}{1 \cdot 2 \cdots (n-1)n} = \frac{2^{2n-1}}{n} = \frac{2^{2n}}{2n}.$$

Considering the two bounds together, we get

$$(2n)^{\pi(2n)} > \frac{2^{2n}}{2n} \qquad \text{and therefore} \qquad \pi(2n) + 1 > \log 2 \frac{2n}{\log 2n}.$$

With this we now have the desired bound for all even integers, and with a little additional calculation, we can get a similar result for all x:

$$\pi(x) \geq \pi\left(2\left[\frac{x}{2}\right]\right) > \log 2 \frac{2\left(\frac{x}{2}-1\right)}{\log x} - 1 = \log 2 \frac{x}{\log x} - \left(\frac{2\log 2}{\log x} + 1\right).$$

The *subtrahend* (the quantity that is subtracted) is bounded for $x \geq 2$, and the following theorem holds:

Theorem 5 *For every x large enough, there exists a positive constant c for which the following inequality holds:*[15]

$$\pi(x) > \frac{cx}{\log x}.$$

11. We will deduce an upper bound of an estimate for the product of all primes less than a number x, which is interesting in itself.

Theorem 6 *For the product of the primes not greater than x, we have*

$$\prod_{p \leq x} p \leq 4^x.$$

It is enough to prove the theorem for integers x, for if it is true for integers, then for arbitrary x we have

---

[15] For all the details of the above proof, see E. LANDAU: *Vorlesungen über Zahlentheorie* (1927), Volume 1, p. 67.

$$\prod_{p \leq x} p = \prod_{p \leq [x]} p \leq 4^{[x]} \leq 4^x.$$

We can prove the theorem for integers x by induction. The theorem is obviously true for $x = 1$ and $x = 2$. Let $k \geq 3$ be an integer. Assume that the theorem is true for all positive integers less than k, and we will show that the theorem is true for $x = k$. This is immediately true if k is even, since the assumption that $k \geq 3$ implies that

$$\prod_{p \leq x} p = \prod_{p \leq x-1} p \leq 4^{x-1} \leq 4^x.$$

Let k be odd, i.e., $k = 2n + 1$. Then we can split the product into two parts. We apply the induction hypothesis to estimate the part of the product made up of primes not greater than $n + 1$; we will give a separate bound for the other part.

The primes between $n+2$ and $2n+1$ (including the endpoints) are divisors of the binomial coefficient $\binom{2n+1}{n}$, since this is an integer, and in the form

$$\binom{2n+1}{n} = \frac{(n+2)(n+3)\cdots(2n+1)}{1 \cdot 2 \cdots n},$$

these primes occur in the numerator and are not divisors of the denominator, since every factor is smaller than the primes. Thus the product of the primes between $n + 2$ and $2n + 1$ is not larger than this binomial coefficient, and it is enough to give a bound for this from above.

This can be estimated similarly to the binomial coefficient $\binom{2n}{n}$, only now we need an upper bound, so we replace each odd number in the numerator by the next-larger even number. The binomial coefficient

$$\binom{2n+1}{n} = \frac{2 \cdot 3 \cdot 4 \cdots 2n \cdot (2n+1)}{(1 \cdot 2 \cdot 3 \cdots n)(n+1)!}$$

can be simplified by first dividing the product of the even integers in the numerator by $n!$, and then increasing all remaining factors by 1 to get

$$\binom{2n+1}{n} = 2^n \cdot \frac{3 \cdot 5 \cdot 7 \cdots (2n+1)}{2 \cdot 3 \cdots (n+1)} < 2^n \cdot \frac{4 \cdot 6 \cdots (2n+2)}{2 \cdot 3 \cdots (n+1)} = 4^n.$$

Using the inductive hypothesis and the above inequality, we have

$$\prod_{p \leq 2n+1} p = \prod_{p \leq n+1} p \cdot \prod_{n+2 \leq p \leq 2n+1} p < 4^{n+1} \cdot 4^n = 4^{2n+1}.$$

With this we have proved the theorem.

12. The number of primes here is the number of factors. If we wish to bring this into the inequality, using the fact that each prime is at least 2, we arrive at an inequality that tells us nothing. We can greatly improve upon this by replacing all primes greater than some large number y by the smallest of those. By this step, however, the number of factors is also diminished:

$$\prod_{y \leq p \leq x} p > y^{\pi(x) - \pi(y)}.$$

We would like to get a bound of the order $x / \log x$ for $\pi(x)$. The subtrahend in the exponent will be smaller than this if we choose y to be x to some power δ, where δ is less than 1. We may then estimate the subtrahend $\pi(y)$ in the exponent by y, and substituting we get

$$y^{\pi(x) - \pi(y)} > x^{\delta\left(\pi(x) - x^\delta\right)}.$$

Multiplying this by the remaining primes, the product will still be smaller than 4^x. From here we can get the desired inequality by omitting the primes less than y (we estimate their product from below by 1). Using Theorem 6 we get that if δ is less than 1, then

$$4^x > \prod_{y \leq p \leq x} p > x^{\delta\left(\pi(x) - x^\delta\right)}.$$

Taking logarithms of the outer parts and solving for $\pi(x)$, we have

$$\pi(x) < \frac{\log 4}{\delta} \cdot \frac{x}{\log x} + x^\delta = \left(\frac{\log 4}{\delta} + \frac{\log x}{x^{1-\delta}}\right) \frac{x}{\log x}.$$

Here the second term in parentheses is bounded for $\delta < 1$ by Fact 7, and in fact, it can be made arbitrarily close to 0 for x large enough. With this we have the following theorem:

Theorem 7 *There exists a positive constant c for which the following inequality holds:*[16]

$$\pi(x) < \frac{cx}{\log x}.$$

The present line of thought is essentially due to ERDŐS and KALMÁR.[17]

If we interpret Theorem 3 to mean that the prime numbers cannot occur too infrequently among the natural numbers, then Theorem 7 means that the

---

[16] From the proof, we will get that any number greater than $\log 4$ can be used as c, for instance 1.4, for all x larger than some (determinable) bound. The value of the constant, however, is secondary, since any value greater than 1 will suffice for x large enough, but the proof of this fact depends on considerably deeper investigations.

[17] P. ERDŐS: *Acta Litt. Univ. Sci., Szeged, Sect. Math.* **5** (1932), pp. 194–198.

primes are not too dense either. This is because if we ask the question that up to a limit x, what fraction of natural numbers are prime, then according to Theorem 7 the answer is $\pi(x)/x$, which in turn is less than $c/\log x$, which is arbitrarily small as x grows sufficiently large.

If the quotient of the number of elements of a sequence of integers not greater than a given number x divided by x is arbitrarily small for x large enough, then we frequently say that the sequence *has density* 0 (or its asymptotic density is 0). Therefore, the sequence of primes has density 0. This weaker statement can, however, be proved using simpler techniques, without referring to Theorem 7 (see Exercise 10).

13. We note that if in addition to the primes, we include the prime powers too, then the situation does not change considerably. Let $\pi^*(x)$ denote the number of all prime powers up to x. Then the following theorem holds:

Theorem 8 *There are positive constants C_1 and C_2 such that the number $\pi^*(x)$ of all prime powers not larger than x satisfies the inequalities*

$$\frac{C_1 x}{\log x} < \pi^*(x) < \frac{C_2 x}{\log x},$$

for x large enough.

The first inequality follows from Theorem 5:

$$\pi^*(x) \geq \pi(x) > \frac{cx}{\log x},$$

so choosing $C_1 = c$ from Theorem 5 satisfies the first inequality.

For the upper bound, we see that up to x, only the squares of those primes occur that are not larger than $\sqrt{x}$, and only the cubes of the primes up to $\sqrt[3]{x}$, etc. Thus

$$\pi^*(x) = \pi(x) + \pi\left(\sqrt{x}\right) + \pi\left(\sqrt[3]{x}\right) + \cdots.$$

On the right-hand side there are only finitely many terms. The last term appearing is the kth root, where k satisfies

$$\sqrt[k]{x} \geq 2 > \sqrt[k+1]{x}.$$

From the first inequality, we have

$$k \leq \frac{\log x}{\log 2}.$$

Because $\pi(x)$ is nondecreasing in x, using the c from Theorem 7, it follows that

$$\pi^*(x) \le \pi(x) + (k-1)\pi(\sqrt{x}) < \frac{cx}{\log x} + \frac{kc\sqrt{x}}{\log\sqrt{x}}$$

$$< \frac{cx}{\log x} + \frac{2c}{\log 2}\log x \frac{\sqrt{x}}{\log x} = \frac{cx}{\log x} + \frac{2c}{\log 2}\sqrt{x} < \frac{C_2 x}{\log x},$$

for an appropriate constant C_2. With this we have proved Theorem 8.

Exercises:

8*. Beyond what bounds on x do the following inequalities hold?
 (a) $\pi(x) \le x/2$,
 (b) $\pi(x) \le (x-1)/2$,
 (c) $\pi(x) \le x/3$.

9. Let $0 < a_1 < a_2 < \cdots$ be a sequence of natural numbers, and denote by $A(x)$ the number of elements not greater than x. Prove that if for every k there exists a c_k such that

$$A(x) > (\log x)^k$$

holds whenever $x > c_k$, then there are infinitely many distinct primes among the prime divisors of the elements of the sequence.

10. Let the elements of the sequence of natural numbers $a_1, a_2, \ldots,$ be pairwise relatively prime. Let $A(x)$ be the number of elements less than x. Find a bound for $A(x)$ using the fact that $a_1, a_2, \ldots, a_k$, do not divide any other elements of the sequence. Prove that the sequence has density 0.

11*. Prove that there exists a prime between any n and 2^n.

12. Prove that

$$\frac{2^{2n}}{2\sqrt{n}} < \binom{2n}{n} < \frac{2^{2n}}{\sqrt{2n+1}}.$$

14. The proofs of Theorems 6 and 7 were founded on estimates of the binomial coefficients $\binom{2n}{n}$ and $\binom{2n+1}{n}$, respectively. More carefully estimating the binomial coefficient, we are able to prove the following theorem:

Theorem 9 *For every positive integer n, there is a prime number p satisfying*

$$n < p \le 2n. \tag{1}$$

BERTRAND was the first to observe this fact, and he even used it, but it was CHEBYSHEV who first proved it somewhat later, in 1850.[18]

---

[18] J. BERTRAND needed only the existence of a prime between n and $2n-2$ when $n \ge 4$. P. L. CHEBYSHEV proved the existence of such a prime in an even smaller interval.

The product of primes between $n+1$ and $2n$, if there are any, divides the binomial coefficient

$$\binom{2n}{n} = \frac{(n+1)(n+2)\cdots 2n}{1 \cdot 2 \cdots n}, \qquad (2)$$

since all these primes divide the numerator, but none divides the denominator. We will show that this product of primes has at least one element. We have already seen in Section 10 that the binomial coefficient is larger than $2^{2^n}/2n$.

To get a lower bound for the product of the primes above, we need to get an upper bound on the product of powers of primes not greater than n. If we estimate each from above by $2n$, nothing useful comes of it. It is necessary to estimate the product of those primes that appear to at most the first power by the result of Theorem 6. These are those primes from Theorem 1.12 that occur to the power 1, i.e., $p^2 > 2n$ and $p > \sqrt{n}$. Thus the binomial coefficient can be decomposed into the following three products:

$$T_1 = \prod_{p < \sqrt{2n}} p^{\beta(p)}, \quad T_2 = \prod_{\sqrt{2n} \le p \le n} p^{\beta(p)}, \quad T_3 = \prod_{n+1 \le p \le 2n} p.$$

We want to estimate T_3 from below; to do this we need to estimate both T_1 and T_2 from above.

In T_2, every exponent is either 0 or 1. If every exponent is 1, then disregarding the smaller primes that are missing in the product, the bound of 4^x from Theorem 6 cannot be significantly improved, and it is almost as large as the lower bound for the binomial coefficient. The important observation is that at the end of the product, the exponents of the primes larger than $\frac{2}{3}n$ are 0.

If $\frac{2}{3}n < p \le n$, then p appears in the denominator of (2), but $2p$ does not. In the numerator, $2p$ appears, whereas $3p > 2n$ does not. If $n > 2$, then $p > 2$ and $p \ne p^2$, so after simplifying, p cancels.

Estimating the factors of T_1 above by $2n$, we get

$$\frac{2^{2n}}{2n} \le \binom{2n}{n} = T_1 T_2 T_3 < (2n)^{\pi(\sqrt{2n})} 4^{2n/3} T_3,$$

and this gives a lower bound for T_3:

$$T_3 > \frac{2^{2n/3}}{(2n)^{\pi(\sqrt{2n})+1}}.$$

We can further decrease this by increasing $\pi\left(\sqrt{2n}\right)$. The prime numbers, with the exception of 2, are all odd. Among the odds, 1 is not a prime number, and if $\sqrt{2n} > 15$, then up to $\sqrt{2n}$, neither is 9 nor 15. Thus for such n, we have

$$\pi\left(\sqrt{2n}\right) + 1 \le \left(\frac{\sqrt{2n}+1}{2} - 2\right) + 1 = \frac{\sqrt{2n}-1}{2} < \frac{\sqrt{2n}}{2}.$$

Using additionally that if $\sqrt{2n} > 15$ then $n \ge 113$, we obtain

$$T_3 > \frac{2^{2n/3}}{(2n)^{\sqrt{2n}/2}} = \frac{2^{2n/3}}{\left(\sqrt{2n}\right)^{\sqrt{2n}}} = \left(\frac{2^{\sqrt{2n}}}{\left(\sqrt{2n}\right)^3}\right)^{\sqrt{2n}/3}.$$

The right-hand side is larger than 1 if the fraction in parentheses is larger than 1, at least for large enough n. Denoting $\sqrt{2n}$ by z, we need to show that $2^z > z^3$ for z above some bound. It is enough to prove that

$$2^m > (m+1)^3 \tag{3}$$

for integers m, for if this is true, then for arbitrary positive real numbers z,

$$2^z \ge 2^{[z]} \ge ([z]+1)^3 > z^3.$$

The inequality in (3) is true for $m = 11$ ($2^{11} = 2048, 12^3 = 1728$). For larger values of m, the reader can easily prove the inequality by induction.

The fraction under investigation, then, is greater than 1 whenever $\sqrt{2n} > 11$, which is true for every n at least 61. The earlier inequality was valid for all n larger than 113, so to complete the theorem, we need to check the numbers up to 113. It is enough to give a list of primes in increasing order such that each prime is less than twice the previous prime, for example, the primes

$$2, 3, 5, 7, 13, 23, 43, 83, 127.$$

For $n = 1$, the inequalities (2) of the theorem are satisfied by 2 (this is why we include the upper bound). For larger values of n,

$$p \le n \le p',$$

where p and p' are consecutive elements of the list of primes, we see that

$$p' < 2p \le 2n,$$

and that p' satisfies (1). This completes the proof of Theorem 9.

15. Theorem 9 can also be phrased (and it has also been proven in this form) that the binomial coefficient $\binom{2n}{n}$ has a prime divisor greater than n. The theorem in this form can be essentially improved to get the following theorem due to SYLVESTER and SCHUR.

Theorem 9′ *If $n \ge 2k$, then the binomial coefficient $\binom{n}{k}$ has a prime divisor greater than k.*

However, even the simplest, shortest proof to this theorem (due to ERDŐS) is too long for us to discuss here.[19]

Exercises:

13. Prove the inequality in (3).

14*. Assuming[20] that there is always a prime between n and $2n - 7$ for $n \geq 10$, prove that, with the exception of 1, 4, and 6, every natural number can be written as the sum of distinct primes.

15*. Give another proof of the fact that if m and n are positive integers, then
$$\frac{1}{n} + \frac{1}{n+1} + \cdots + \frac{1}{n+m}$$
is not an integer.

16*. Prove that for every polynomial with integer coefficients of degree at least one, there are infinitely many integers for which the value is composite.

16. Let us now review the proofs of this chapter. In Section 5, decomposing integers into a square and a square-free part, we arrived at the inequality $x \leq \sqrt{x} 2^{\pi(x)}$. In the indirect proof in Section 7 we arrived at a similar inequality.

In Section 9 we saw that
$$\prod_{p \leq x} \left(1 - \frac{1}{p}\right)^{-1} > \sum_{n \leq x} \frac{1}{n}.$$

In Section 10, using the inequality
$$\prod_{n+2 \leq p \leq 2n+1} p < \binom{2n+1}{n}$$
we were able to derive the inequality
$$\prod_{p \leq x} p < 4^x.$$

In Section 12 we used the inequality
$$\binom{2n}{n} \leq (2n)^{\pi(2n)}.$$

Finally, in Section 14, by more closely investigating the prime decomposition of the binomial coefficient $\binom{2n}{n}$, we arrived at the inequality

---

[19] P. ERDŐS: *Journ. London Math. Soc.* **9** (1934), pp. 282–288.
[20] The reader can prove this by applying the proof of Theorem 9 with slight modifications to the binomial coefficient $\binom{2n-6}{n-6}$.

$$\prod_{n+1\le p<2n} p > \left(\frac{2^{\sqrt{2n}}}{\left(\sqrt{2n}\right)^3} \right)^{\sqrt{2n}/3} .$$

Each of these inequalities involves an expression independent of the prime numbers, namely,

$$[x], \quad \sum_{n\le x}\frac{1}{n}, \quad \binom{2n}{n}, \quad 4^x, \quad \binom{2n+1}{n}, \quad \left(\frac{2^{\sqrt{2n}}}{\left(\sqrt{2n}\right)^3} \right)^{\sqrt{2n}/3} ,$$

and an expression depending on the primes, and from these we were able to learn more about the primes. Every expression of this type promises insights into the properties of the primes.

Up to now we have succeeded in showing that the order of magnitude of $\pi(x)$ is $x/\log x$, in other words, that $\pi(x)/(x/\log x)$ is between positive bounds. Experience suggests that more is true: $\pi(x)$ is "essentially" $x/\log x$ for large values of x, which says that the ratio above is not just between positive bounds, but that it is arbitrarily close to 1 for x large enough. We say that the *asymptotic value*[21] of $\pi(x)$ is $x/\log x$.

This is known as the *prime number theorem* and is much more difficult to prove than what we have proved so far. B. RIEMANN (1826–1866) outlined a proof in an ingenious work using tools from complex analysis.[22] His work contained many statements without proof. It was first in 1896 that his ideas were further developed into a complete proof of the theorem. This was done independently by HADAMARD[23] and DE LA VALLÉE-POUSSIN.[24] We will not try to outline the proof here, but we will, however, mention some of the identities upon which the proof builds.

In the elements of analysis it is shown that the infinite series

$$1 + \frac{1}{2^s} + \frac{1}{3^s} + \cdots$$

converges for all real numbers s greater than 1. In fact, this is also true for all complex numbers s with real part greater than 1. Let $\zeta(s)$ denote its value (depending on s).

Following in a way similar to that of Sections 8 and 9, it can be seen that in this domain the following relation holds:

$$\zeta(s) = \prod_p \left(1 - \frac{1}{p^s}\right)^{-1} .$$

---

[21] The so-called *integral logarithm* $\mathrm{Li}(x) = \int_2^x (1/\log t)dt$ describes the behavior of $\pi(x)$ with an even smaller deviation. Its asymptotic value is also $x/\log x$.

[22] B. RIEMANN: *Monatsberichte d. Berliner Acad. d. Wiss.* (1859), pp. 671–680.

[23] J. HADAMARD: *Bulletin Soc. Math. France* **24** (1896), pp. 199–220.

[24] CH. DE LA VALLÉE-POUSSIN: *Annales Soc. Sci. Bruxelles* **20** (1896), pp. 183–256 and pp. 281–297.

Again we have a relation between a function having nothing to do with the primes and a function depending on the primes. This time the relation is equality.

The definition of $\zeta(s)$ can be extended to all complex numbers s, with the exception of 1. The investigation of this function and the above identity leads to a proof of the prime number theorem.

In the middle of the twentieth century the theorem was finally proved without the use of analysis.[25]

Comparing the prime number theorem and Chebyshev-type theorems, one can see that it follows from Theorem 9 that

$$\frac{p_{n+1}}{p_n} < 2,$$

and from the prime number theorem it follows that this ratio is asymptotically 1. It is not likely that this last statement could be proved with basic ideas like those that we have used here.

In Section 2.19 we stated DIRICHLET's famous theorem: *"If a and b are relatively prime integers, a positive, then there are infinitely many primes in the arithmetic progression $ak + b$ $(k = 1, 2, \ldots)$."*

We have also shown some interesting applications of this theorem. Obviously, it is enough to examine those arithmetic progressions in which the value of b is between 1 and a. Among these, $\varphi(a)$ integers are relatively prime to a. It has been shown that the primes are distributed evenly within these $\varphi(a)$ arithmetic progressions, in the sense that if we denote by $\pi(a, b, x)$ the number of primes not greater than x in $ak + b$, then the asymptotic value[26] of $\pi(a, b, x)$ is $x/\varphi(a) \log x$, as long as $(a, b) = 1$ holds.

Previously, it has been mentioned that number-theoretic notions can be extended to other areas besides the integers. In numerous cases it has been possible to prove analogous theorems. One such theorem can be stated that remains in the domain of integers. Let $f(x)$ be a polynomial with integer coefficients that cannot be written as the product of lower-degree polynomials with rational coefficients. Additionally, let p be a prime. Denote by $\phi(p)$ the number of integers n between 0 and $p - 1$ for which $f(n)$ is divisible by p. Then

$$\frac{\sum_{p \leq x} \phi(p)}{x/\log x}$$

is arbitrarily close to 1 for x sufficiently large. (A special case of this is the prime number theorem, namely, when $f(x) = x$.)

---

[25] P. ERDŐS: *Proceedings Nat. Acad. Sci.* **35** (1949), pp. 374–384. A. SELBERG: *Annals of Math.* **50** (1949), pp. 305–313. For a detailed form of the first paper see GY. HOFFMANN, L. SURÁNYI: *Matematikai Lapok* **23** (1972), pp. 31–51 (in Hungarian).

[26] For more details, see the second part of CH. DE LA VALLÉE-POUSSIN's paper from footnote 8.

17. Simple proofs have been found for some special cases of Dirichlet's theorem, mentioned above. For instance, it is obvious that there are infinitely many primes of the form $2k + 1$ (i.e., odd), since with the exception of 2, every prime is of this form.

With respect to 4, the odd primes can be of the forms $4k + 1$ and $4k - 1$. Similarly, with respect to 6, primes other than 2 and 3 fall into two classes, either of the form $6k + 1$ or $6k - 1$. We now prove that there are infinitely many primes in each of these classes.

Theorem 10 *There are infinitely many positive integers k such that:*

(a) $4k + 1$ *is prime,*
(b) $4k - 1$ *is prime,*
(c) $6k + 1$ *is prime,*
(d) $6k - 1$ *is prime.*

All the proofs use the basic Euclidean idea that we used to prove the existence of infinitely many primes, except that we are now looking for expressions that have a prime divisor of the required form. This is easier for the cases (b) and (d).

In general, if $a > 2$, and a number is of the form $an - 1$ (in other words, it is congruent to -1 modulo a), then it has a prime divisor that is not congruent to 1 modulo a. This is quite clear, since the product of numbers congruent to 1 modulo a is also congruent to 1 modulo a. If $a = 4$ or $a = 6$ (the cases in question above), then there are only two relatively prime residue classes modulo a. Hence, in these cases, if a number relatively prime to a is not congruent to 1 modulo a, then it is congruent to -1 modulo a.

Therefore, if $a = 4$ or $a = 6$, then 3 and 5 are primes of the form $ak - 1$ ($k = 1$ in both cases). On the other hand, if $p_1, p_2, \ldots, p_r$ are given primes of the form $ak - 1$, then there are other primes of this form, too, that are not in the above list. For instance, the number

$$n = ap_1 p_2 \cdots p_r - 1$$

has a prime divisor of the form $ak - 1$, and this is different from all the p_i's ($i = 1, 2, \ldots, r$), since dividing n by p_i yields the remainder $p_i - 1$.

18. To prove parts (a) and (c), we will use two results from Chapter 2. The first is an application of Euler's lemma (Theorem 2.21) seen in Section 42: If a is an integer, then every odd prime divisor of $a^2 + 1$ is of the form $4k + 1$. The other result is an application of Gauss's lemma (Theorem 2.22 from Section 2.44): Prime divisors different from 2 and 3 of numbers of the form $a^2 + 3$ must be of the form $6k + 1$. Using these we are able to prove the remaining parts of Theorem 10:

(a) $5 = 4 \cdot 1 + 1$, a prime of the type we are looking for. If $p_1, p_2, \ldots, p_r$ are primes of the form $4k + 1$, then every prime divisor of

$$(2p_1 p_2 \cdots p_r)^2 + 1$$

is of the type $4k + 1$, different from those above.

(c) $7 = 6 \cdot 1 + 1$ is a prime of the type we are looking for. If $p_1, p_2, \ldots, p_r$ are primes of the form $6k + 1$, then every prime divisor of

$$(2p_1 p_2 \cdots p_r)^2 + 3$$

is of the type $6k + 1$, different from those above, since this new number is not divisible by either 2 or 3.

With this we have proved all the claims of Theorem 10.

19. It may seem of excessively precise that we have begun our proofs by first exhibiting a prime of the desired form. One may reduce the proof of the general theorem to proving the existence of one prime of the desired form. That is because if it is true for all relatively prime integers a and b, a different from 2, that there exists a positive integer k such that $ak + b$ is a prime, then the theorem could be applied to all of the arithmetic progressions

$$ak + b, a^2 k + b, a^3 k + b, \ldots \ (k = 1, 2, \ldots),$$

since a^c and b are also relatively prime. The primes in these sequences are all of the form $ak + b$, and there are arbitrarily large ones among them, too, so therefore there are infinitely many different primes of the given type as well.

20. Certain other cases of Dirichlet's theorem have been proven by elementary methods; in fact, elementary proofs have been found for general classes, such as showing that there are infinitely many primes of the form $ak + 1$, where a is an integer larger than 2. M. BRAUER found a rather simple proof of the fact that there are infinitely many primes of the form $ak + 1$ and $ak - 1$, for every positive integer a. Many notions in the proofs of the two cases run parallel to each other. Below we will prove only the case of $ak + 1$, and then only for a a prime power.

Theorem 11 *For every positive prime p and every positive integer c, there are infinitely many primes of the form $p^c k + 1$.*

It is enough in this case, too, to show the existence of one prime of the given form, for if there exists at least one prime of the given type for every c, then replacing c by $c' = c + m$, for $m = 1, 2, \ldots$, in the theorem, we have primes of the form $p^{c+m} k + 1 = p^c (p^m k) + 1$, and among these there are arbitrarily large ones, and hence there are infinitely many.

21. We are looking for a prime q that satisfies $p^c \mid q - 1$. Based on this observation, Fermat's theorem and the idea of the order of a number (modulo q) suggests a plan for a proof. Fermat's theorem states that if a prime q is not a divisor of an integer u, then $u^{q-1} \equiv 1 \pmod{q}$.

The order of u is the smallest positive integer r for which

$$u^r \equiv 1 \pmod{q}.$$

According to Theorem 2.17, if

$$u^s \equiv 1 \pmod{q}, \qquad \text{then} \qquad r \mid s.$$

Based on this, we need to find a prime q and an integer u not divisible by q for which

$$u^{p^c} \equiv 1 \pmod{q}, \tag{4}$$

and the power of u is its order modulo q. If these are satisfied, then by Fermat's theorem and the property of the order of a number, we have that $p^c \mid q - 1$, meaning that q is of the desired form.

If a prime q and an integer u satisfy (4), then the order of u is a divisor of p^c; hence it is either equal to it or it is p raised to a power smaller than c. In the latter case, the order is a divisor of p^{c-1} as well, and hence the congruence

$$u^{p^{c-1}} \equiv 1 \pmod{q}$$

is satisfied. This we would like to rule out.

Thus we have reduced the proof of the theorem to the following task: We are looking for a prime q different from p and an integer u such that q is a divisor of $u^{p^c} - 1$, but not a divisor of $u^{p^{c-1}} - 1$.

We claim that for arbitrary u, every prime divisor of

$$f(u) = \frac{u^{p^c} - 1}{u^{p^{c-1}} - 1}$$

different from p (if such a divisor exists) satisfies the desired conditions. This is true because on the one hand,

$$u^{p^c} - 1 = f(u)\left(u^{p^{c-1}} - 1\right),$$

and thus the prime divisors of $f(u)$ are also prime divisors of $u^{p^c} - 1$. On the other hand,

$$f(u) = u^{(p-1)p^{c-1}} + u^{(p-2)p^{c-1}} + \cdots + u^{p^{c-1}} + 1$$
$$= \left(u^{(p-1)p^{c-1}} - 1\right) + \left(u^{(p-2)p^{c-1}} - 1\right) + \cdots + \left(u^{p^{c-1}} - 1\right) + p.$$

The differences in parentheses are all divisible by $u^{p^{c-1}} - 1$. Therefore, if a prime q divides both $f(u)$ and $u^{p^{c-1}} - 1$, then it divides p as well; hence it is p. Thus a prime different from p cannot divide both $f(u)$ and $u^{p^{c-1}} - 1$, as we claimed.

Finally, $f(p)$ is larger than 1, and dividing by p gives remainder 1. Thus it has prime divisors, and every one is of the form $p^c k + 1$. With this we have proved Theorem 11.

We note that in the case $p = c = 2$ we have $f(u) = u^2 + 1$, and thus the proof of Theorem 10, Part (a).

22. If the distinct prime divisors of a number a are $p_1, p_2, \ldots, p_r$, then a line of thought similar to that of the previous section leads to the search for a prime q and an integer u for which it is true that q is a divisor of $u^a - 1$ but not a divisor of any of the $u^{a/p_j} - 1$'s, for any j. As we did in the case above we can produce a seemingly rational function similar to f made up of expressions of the form $g_n = u^n - 1$. It can be seen that this function is indeed a polynomial of u with the desired properties. To prove these statements, it is necessary only to use the relation that if $m \mid n$, then $g_m \mid g_n$. We will, however, not go into the details of proving these statements.

We only mention the basic idea of the proof of the statement concerning primes of the form $ak - 1$. M. BAUER noticed that—denoting the imaginary unit by the usual i ($i^2 = -1$)—just as for $g_n(u)$, it is also true for the polynomial $h_n(u) = ((x + i)^n - (x - i)^n)/2i$ with integer coefficients that if $m \mid n$, then $h_m(u) \mid h_n(u)$. And it is possible using the h_n's to build an expression that has a prime divisor of the form $ak - 1$.

He also needed to use the theorem (also due to him, and very interesting on its own) that *if a polynomial with integer coefficients has a real root, then there are infinitely many primes that are divisors of the value of the polynomial at some integral value (other than at places that are congruent to 1 modulo m).*[27] However, all of these require so many technical details that they are beyond the scope of this book.

---

[27] The proofs of these two statements in the general case can both be found in E. LANDAU: *Handbuch der Lehre von der Verteilung der Primzahlen*, 1909, B. G. Teubner, pp. 436–446.

6. Sequences of Integers

1. Certain sequences of numbers, as well as problems relating to them, appear throughout mathematics. Determining the general term of an arithmetic or geometric progression, as well as determining the sum of (finitely) many consecutive terms of these sequences, are common high school problems. A residue class for a given modulus also constitutes an arithmetic progression. In Chapter 2 various problems relating to both disjoint systems of congruences as well as covering systems of congruences for different moduli were discussed; many of these are still unsolved.

The sequence of prime numbers is especially interesting from the number-theoretic point of view. The entire previous chapter was devoted to certain problems relating to the primes. Among other things, we showed that the sequence of primes is infinite, and after introducing the concept of a sequence of 0 density, we proved that the primes have density 0. In Sections 5.3–5.4 we used various sequences of integers to prove that there are infinitely many primes. We also saw various consequences of DIRICHLET's deep theorem about primes in arithmetic sequences. (We proved the theorem only for various special cases.)

In this chapter we will further investigate sequences of integers, considering them from different points of view.

2. In this section we will examine the length of the Euclidean algorithm previously discussed in Section 1.27, and from our investigation we will come upon a well-known sequence. This procedure is used for finding the distinguished common divisor of two integers a and b, and consists of consecutive divisions with remainder. In a footnote we mentioned that a and b can be taken as the -1st and the 0th remainders in the sequence of remainders, r_{-1} and r_0. Suppose that $0 < a < b$ and that $a \nmid b$. Then this algorithm can be briefly described as

$$r_{j-1} = r_j q_j + r_{j+1}, \qquad 0 < r_{j+1} < r_j, \qquad j = 0, 1, \ldots, n - 1,$$
$$r_{n-1} = r_n q_n.$$

Here, the greatest (distinguished) common divisor of a and b is r_n. Let us call the length of the algorithm n.

With the assumptions on the remainders, a and b uniquely determine the sequence of r's and q's. The converse of this is also true: If $q_0, q_1, \ldots, q_n$ and r_n are known, then reversing our steps, these numbers in turn determine all the r_j's, and finally b and a. For a given n, let us determine the smallest a and b for which the length of the Euclidean algorithm is n.

In order to do this, we will show that if for some a' and b', the length of the algorithm is n, and if we denote by $q'_0, q'_1, \ldots, q'_n$ the quotients and by $r'_{-1}, r'_0, r'_1, \ldots, r'_n$ the remainders, if the inequalities $q'_j \geq q_j$, $j = 0, 1, \ldots n$, and $r'_n \geq r_n$ hold, then for $j = n-1, n-2, \ldots 1, 0, -1$ it also holds that

$$r'_j \geq r_j, \quad \text{and especially} \quad a' \geq a, \quad b' \geq b. \qquad (1)$$

We prove this by induction on n, starting from $n-1$, going in descending order. For $i = n-1$, it follows from $r'_n \geq r_n$ and $q'_n \geq q_n$, using the last inequality, that

$$r'_{n-1} = r'_n q'_n \geq r_n q_n = r_{n-1}.$$

If for some nonnegative integer k, (1) holds for $j \geq k$, then since $q'_k \geq q_k$, we see that

$$r'_{k-1} = r'_k q'_k + r'_{k+1} \geq r_k q_k + r_{k+1} = r_{k-1}.$$

Therefore, (1) also holds for $j = k-1$. With this we have shown that (1) holds for all j's.[1]

We also note that the value of q_n is at least 2, since if $q_n = 1$, then r_{n-1} would equal r_n, meaning that at the second-to-last division, $q_{n-1} + 1$ would be the proper quotient, and the algorithm would finish there.

From the statement proved above, it follows that among the set of all pairs of integers for which the algorithm has length n, the smallest pair will be the one for which

$$q_0 = q_1 = q_2 = \cdots = q_{n-1} = r_n = 1, \quad q_n = 2$$

holds.

Note that if for two distinct values of n, we write the smallest pair of integers for which the algorithm length is n, then in performing the algorithm for the larger n, the values of the algorithm for the smaller n appear as neighboring remainders, and from here the two algorithms coincide. In this way we have produced an infinite sequence of remainders, whose finite pieces provide the remainders of the longest such algorithm in question. Hence, it is appropriate to index the remainders in reverse order:

$$r^*_1 = r_n, \quad r^*_2 = r_{n-1}, \quad r^*_3 = r_{n-2}, \quad \ldots.$$

---

[1] It is only a formal difference from the usual way of induction that we proceed in descending order on the indices. It is more unusual that we prove a statement by induction that holds only for finitely many integers, namely only for as many as the number of remainders in the algorithm.

Then for the sequence in question we have that $r_1^* = r_n = 1$. Furthermore, since in the last equality $q_n = 2$,

$$r_2^* = r_{n-1} = r_n q_n = 2,$$

while the remaining equations of the algorithm are

$$r_{j+2}^* = r_{j+1}^* + r_j^* \qquad (j = 1, 2, \dots).$$

If we add the terms 0 and 1 to the beginning of the sequence, the sequence we get,[2] defined by the recurrence relation

$$f_0 = 0, \ f_1 = 1, \ f_{j+2} = f_{j+1} + f_j \ \ (j = 0, 1, 2, \dots) \tag{2},$$

is called the *Fibonacci sequence*, while its elements are called the *Fibonacci numbers*. The first few elements of this sequence are

$$0, \ 1, \ 1, \ 2, \ 3, \ 5, \ 8, \ 13, \ 21, \ 34, \ 55,$$

$$89, \ 144, \ 233, \ 377, \ 610, \ 987, \ 1597, \ \dots$$

We will call other similar sequences that start from different values but satisfy the recurrence relation in (2) *Fibonacci-type sequences*.

LEONARDO FIBONACCI (son of BONACCIO, also called LEONARDO PISANO (FIBONACCI), lived between circa 1180 and 1250), famous Italian mathematician, introduced the European world to the above notion in his book titled *Liber Abaci*, published in 1202, based on his experiences in Arab lands.

He introduced the Fibonacci numbers in his book using a slightly farfetched example concerning the reproduction of rabbits. These numbers proved very interesting and had many applications, and even today, 800 years later, they are the center of intense research. Many problems related to Fibonacci numbers are still unsolved today.

In the above, we proved the following theorem:

Theorem 1 *The Euclidean algorithm on a and $b, a > b$, has fewer than n remainders if $b < f_{n+2}$; and also if $a < f_{n+3}$. On the other hand, the Euclidean algorithm on f_{n+3} and f_{n+2} has exactly n remainders, the numbers $f_{n+1}, f_n, \dots, f_2$.*

Among properties of the Fibonacci sequences that we will discuss, many will also hold for Fibonacci-type sequences.

3. The elements of the sequence listed above grow more and more quickly as they get larger. Nevertheless, the sequence "encompasses" every integer, in the sense that every integer has a multiple in the sequence.

---

[2] The recurrence relation also holds for the starting elements. Furthermore, $f_3 = 2$, and from here on we have the same sequence as before.

Theorem 2 *For every integer m, there is a Fibonacci number divisible by m. (In fact, there are infinitely many of these.)*

In the language of congruences, we are looking for an element of the sequence that is congruent to 0, modulo m. This is true for the first element of the sequence (namely 0 itself), and it is therefore enough to show that the sequence modulo m is periodic. This property is true of all Fibonacci-type sequences of integers.

Theorem 3 *Let m be an integer greater than 1. Then the residues modulo m of the elements of every Fibonacci-type sequence of integers is periodic.*

We will first show that the sequence is periodic from some point on. Let $s_0, s_1, \ldots,$ be a Fibonacci-type sequence. Then the consecutive pairs s_i, s_{i+1} of elements can have at most m^2 different values modulo m. Thus among the elements $s_0, s_1, \ldots, s_{m^2}, s_{m^2+1}$, the $m^2 + 1$ pairs of consecutive terms cannot all be different, there must be two consecutive pairs (s_j, s_{j+1}) and (s_k, s_{k+1}), $j < k$, for which

$$s_j \equiv s_k \pmod{m} \quad \text{and} \quad s_{j+1} \equiv s_{k+1} \pmod{m}.$$

Thus by the recursion formula we have

$$s_{j+2} = s_{j+1} + s_j \equiv s_{k+1} + s_k = s_{k+2} \pmod{m},$$

and continuing in this fashion, we have that for every $n \geq j$,

$$s_n \equiv s_{n+k-j} \pmod{m}. \tag{3}$$

Thus the sequence of residues is periodic from some point on.

The periodicity must start from the beginning, because by the recursion, we have for $j > 0$

$$s_{j-1} = s_{j+1} - s_j \equiv s_{k+1} - s_k = s_{k-1} \pmod{m},$$

and we can continue to the beginning of the sequence, thus showing that (3) holds for all nonnegative integers n. With this we have proved Theorem 3.

For the Fibonacci sequence, it follows that (all congruences modulo m)

$$f_{k-j} \equiv f_0 = 0, \ f_{2(k-j)} \equiv f_{k-j} \equiv f_0 = 0, \ \ldots,$$
$$f_{r(k-j)} \equiv f_0 = 0 \ (r = 1, 2, \ldots).$$

With this we have proved Theorem 2.

4. The proof in the previous section shows that the smallest period of a Fibonacci number divisible by m is not larger than $m^2 - 1$. We will see that the first Fibonacci number divisible by m is not larger than $m^2 - 4m + 7$ and $m^2 - 4m + 8$ for m even and odd, respectively (see Exercise 8). In practice, the actual periods seem to be much smaller as m gets larger. It is known that if m is a prime, then f_{m-1} or f_{m+1} is divisible by m for m of the form $10k \pm 1$ or $10k \pm 3$, respectively. Even this is often not the smallest Fibonacci number divisible by a prime.[3]

Exercises:

1. Find simpler formulas for the following sums:
   (a) $S_n = f_1 + f_2 + \cdots + f_n$,
   (b) $T_n = f_1 + f_3 + \cdots + f_{2n-1}$,
   (c) $U_n = f_2 + f_4 + \cdots + f_{2n}$,
   (d) $V_n = f_1^2 + f_2^2 + \cdots + f_n^2$.

2. Prove that at least four and at most five Fibonacci numbers other than the single-digit ones have the same number of digits in decimal notation.

3. Prove that we get a Fibonacci-type sequence if
   (a) we multiply every element of a Fibonacci-type sequence by the same number;
   (b) we add the corresponding elements of two Fibonacci-type sequences.

4. Are there any Fibonacci-type sequences among the
   (a) arithmetic sequences?
   (b) geometric sequences?

5. Give an explicit expression (depending only on n) for the elements of a Fibonacci-type sequence, in particular for the Fibonacci numbers.

6. Assume that the sequence $s_0, s_1, \ldots$ is of Fibonacci type. Express its elements in the form $s_n = u_n s_0 + v_n s_1$, where u_n and v_n are not dependent on the sequence.

7. (Continued.) Prove that the elements of the sequence $S_n = s_n^2$ ($n = 0, 1, \ldots$) satisfy the recursive relation

$$S_{n+3} = uS_{n+2} + vS_{n+1} + wS_n \quad (n \geq 0),$$

   where u, v, and w do not depend on either n or the sequence.

8. Prove that if $m > 3$, then the smallest index of a nonzero Fibonacci number divisible by m is at most $m^2 - 4m + 6$ if m is odd, and $m^2 - 4m + 7$ if m is even.

---

[3] For further reading on the Fibonacci numbers, we refer the reader to N. N. VOROBEV: *The Fibonacci Numbers*, Heath, Boston, 1963. (This is a translation from the Russian.)

5. In Chapter 5 we saw many examples of infinite sequences of pairwise relatively prime elements. In particular, in Sections 3 and 4 we defined some sequences of this type by recursion; in Exercise 5.10 we saw that sequences of this type have density 0, in the sense defined in Section 5.12.

There are other definitions of density of sequences that prove to be useful for certain applications. The previous definition of density 0 is a special case of asymptotic density. It can be defined in the following way: Let $a_1, a_2, \ldots$ be a sequence A and denote by $A(x)$ the number of terms of A not greater than x. We say that the sequence has *upper density* c if the ratio $A(x)/x$ has c as its upper bound. This means that for every $h > 0$, we have $A(x)/x < c+h$ for x large enough, and beyond any bound there is always an N for which $A(N)/N > c - h$. We similarly define the *lower density* of a sequence to be d if it has lower bound d.

If the lower density and the upper density of a sequence coincide, meaning that the sequence $A(x)/x$ has a limit as x gets large, then we say that this limit is the *asymptotic density* of the sequence. It is clear that 0 density is the same as asymptotic density 0.

It follows from the definition that if we know only that for any $h > 0$, the inequality $A(x)/x < c+h$ (respectively $A(x)/x > c - h$) holds for all x large enough, then the upper density is not greater than c (respectively the lower density is at least c).

Exercises:

9. Determine the lower and upper densities of the following sequences:
   (a)   10, 11, 12, ...,
   (b)   1, 4, 9, ...,
   (c) $2 + (n-1)d$ $n = 1, 2, \ldots$,
   (d) $2n + (-1)^n n$ $n = 1, 2, \ldots$.

10*. Find a sequence with upper density 1 and lower density 0.

6. In a large part of what follows we deal with finite sequences. We will see that the infinite analogues of these problems are often significantly more difficult.

We will try to choose as many numbers as possible from the integers between 1 and N in such a way that the pairwise differences are not among the numbers chosen. It is not difficult to see that the odd numbers is such a set, since the difference between any two is even. Another set satisfying this condition is the set of all numbers greater than $N/2$, since the difference between any two is strictly less than $N/2$.

In both examples we have chosen $N/2$ numbers when N is even, and $(N + 1)/2$ when N is odd. We can write this as $[(N + 1)/2]$. It is not too difficult to see that a set of this type cannot be any larger. If n is the largest element of the set $(n \leq N)$, then for all integers k less than n, we may choose only one of k and $n-k$, thus giving an upper bound of $[(n+1)/2] \le [(N+1)/2]$ on the number of elements.

7. Let us now try to partition all integers from 1 to N into sets such that two numbers and their difference do not all appear in the same set.

Corresponding to the first way we chose the integers in the above section, we can partition the integers now such that we pick all the odd integers between 1 and N in the first group, twice the odd integers in the second, four times the odd integers in the third, and so on, until all the numbers are accounted for. This way we get r groups, where r is the integer for which

$$2^r \le N < 2^{r+1}, \qquad \text{i.e.,} \qquad r = \left\lceil \frac{\log N}{\log 2} \right\rceil,$$

holds.

The second method of choosing the integers in the previous section can also be extended to the current problem and produces essentially the same number of sets. However, using a little cleverer methods, one can get fewer sets satisfying the conditions. The question of the least number of sets required naturally arises. Similarly, one can ask whether it is possible to partition *all* the natural numbers into finitely many sets such that no set contains two numbers and their difference. For the first question, we know far less than an exact answer, but we can answer the second question with a definitive "no." More precisely, we prove the following theorem.

Theorem 4 *If the first N natural numbers can be partitioned into r sets such that no two numbers and their difference occur in the same set, then the following relation must hold:*
$$N < r!e.$$

Here e is the base of the natural logarithm, mentioned in Fact 5.

Let us start from a partition of the first N natural numbers into r sets that satisfies the condition. Let the number of elements of a largest set be n_1. Then
$$N \le n_1 \cdot r.$$

Let the elements of one of these largest sets be
$$a_1 < a_2 < a_3 < \cdots < a_{n_1},$$
and call this set the first.

This set cannot contain any of the numbers
$$a_2 - a_1, \ a_3 - a_1, \ \ldots, \ a_{n_1} - a_1,$$
and hence these numbers belong to some of the other $r - 1$ sets. Let n_2 be the largest number of the above elements that belong to the same set. Let these elements be

$$b_1 < b_2 < b_3 < \cdots < b_{n_2},$$

and call the set they belong to the second. Then by the definition of n_2, we have that

$$n_1 - 1 \le n_2(r - 1),$$

and we also have

$$b_i = a_{k_i} - a_1 \quad (i = 1, 2, \ldots, n_2)$$

for appropriate indices $k_1, k_2, \ldots, k_{n_2}$. Again,

$$b_i - b_1 \quad (i = 2, 3, \ldots, n_2)$$

cannot belong to the second set, but they also do not belong to the first set either, since the difference

$$b_i - b_j = a_{k_i} - a_{k_j}$$

is also the difference of two elements from the first set. Therefore, these numbers are distributed among the remaining $r - 2$ sets. If the maximal number of these in the same set is n_3, then

$$n_2 - 1 \le n_3(r - 2).$$

This procedure continues until the $(s - 1)$st step, when the $n_{s-1} - 1$ differences generated by the selected n_{s-1} elements are distributed among the leftover possible $r - s + 1$ sets such that into each set at most $1 = n_s$ elements fall. Then

$$n_{s-1} - 1 \le n_s(r - s + 1) = r - s + 1$$

holds. Obviously, $s \le r$, and for any two elements of the sequence defined this way, $n_1, n_2, \ldots, n_s = 1$, the inequality

$$n_i - 1 \le n_{i+1}(r - i)$$

holds, where $i = 1, 2, \ldots, s - 1$. Additionally, the inequality also holds for $i = 0$ if we interpret n_0 as $N + 1$.

Applying this in turn for all $i = 0, 1, 2, \ldots, s - 1$, and using the expression for e from Section 3.14*, we arrive at the following inequalities:

$$N \le ((n_1 - 1) + 1)r \le n_2(r - 1)r + r \le n_3(r - 2)(r - 1)r + (r - 1)r + r \le \cdots$$
$$\le n_{s-1}(r - s + 2)(r - s + 3) \cdots (r - 1)r + (r - s + 3) \cdots (r - 1)r$$
$$\quad + \cdots + (r - 1)r + r$$
$$\le (r - s + 1)(r - s + 2) \cdots (r - 1)r + (r - s + 2) \cdots (r - 1)r + \cdots$$
$$\quad + (r - 1)r + r$$
$$\le r! + 2 \cdot 3 \cdots (r - 1)r + 3 \cdots (r - 1)r + \cdots + (r - 1)r + r$$
$$= r! \left(1 + \frac{1}{1!} + \frac{1}{2!} + \cdots + \frac{1}{(r - 2)!} + \frac{1}{(r - 1)!} \right) < r!e,$$

which is what we wanted to prove.

8. The theorem and proof given here are due to I. SCHUR. He used them in the proof of the following theorem:

Theorem 5 *For every positive integer m, and for every prime p larger than a bound depending only on m, there are integers $a, b,$ and c, none divisible by p, for which*

$$a^m \equiv b^m + c^m \pmod{p}. \tag{4}$$

This means that a finite-modulus version of Fermat's last theorem fails to be true, so much so that for every m and every p (large enough), there are solutions to (4). (See the following section for a discussion of Fermat's last theorem.)

The proof of this theorem uses the primitive roots of congruence and the notion of index (see Section 2.38) as well as Theorem 2.18, which says that for prime moduli there exist primitive roots.

Let m be a positive integer, p a prime larger than $e \cdot m!$, and g a primitive root modulo p. Partition the positive integers less than p into m classes depending on the residue class modulo m of the index. According to Theorem 4, there is a class having elements $u_1, u_2,$ and u_3 for which $u_1 - u_2 = u_3$, i.e.,

$$u_1 = u_2 + u_3. \tag{5}$$

From the definition of the classes, there is a number i such that every element of this class is congruent modulo p to g^{tm+i}, for some positive integer t. Thus there are integers t_1, t_2, t_3 for which

$$u_j \equiv g^{t_j m + i} \pmod{p}, \quad j = 1, 2, 3.$$

Substituting this in (5) and dividing by g^i, we get that

$$g^{t_1 m} \equiv g^{t_2 m} + g^{t_3 m} \pmod{p}.$$

Thus

$$a = g^{t_1}, \quad b = g^{t_2}, \quad c = g^{t_3}$$

satisfy the congruence (4).

9. *Fermat's last theorem* was written by PIERRE DE FERMAT in the margin of a textbook in the 17th centruy.

Theorem 6 (Fermat's Last Theorem) *The equation*

$$x^m + y^m = z^m$$

has no solutions in positive integers x, y, z for any integer $m > 2$.

FERMAT did not supply a proof of the statement. Instead, he wrote, "I have a truly marvelous demonstration of this proposition which this margin is too narrow to contain." Since the statement was left unproven, it was only a conjecture, and remained as such for 350 years until ANDREW WILES proved it in 1993. It is widely accepted that Fermat did not have a correct proof. However we may never know for certain.

WILES encountered *Fermat's last theorem* as a ten-year-old boy and became fascinated by it. Later, during his studies he became acquainted with the theory of elliptic curves and the theory of modular forms, as well as a conjecture concerning their close relation.[4]

It was clear to Wiles that a proof of this new conjecture would also supply a proof of Fermat's last theorem, but the full proof of this conjectured relation required seven years of hard work. A narrower and narrower gap always seemed to resist his continuing attempts, and he was once at the point of losing all hope of success and giving up, when finally he was able to achieve the last step. In June 1993, he presented a series of three lectures at Cambridge, England, which culminated in the proof of Fermat's last theorem.

10. Let us now return to Theorem 4 and consider partitions of the integers up to N into sets such that no set contains the difference of two of its elements. Experimentation shows that two groups are enough only for the first four integers. Three groups are enough for the first 13; one such partition is the following:

$$\{1, 4, 10, 13\}, \quad \{2, 3, 11, 12\}, \quad \{5, 6, 7, 8, 9\}.$$

The other partitions into three sets differ only in the placement of 7: It can be added to either of the first two sets as well.

We saw in Section 7 that the first 2^r integers can be partitioned into r sets. The partition of the first 13 integers into 3 sets can be easily generalized. It provides a partition of the first $(3^r - 1)/2$ numbers into r sets. As r grows just a little, this is not even close to the upper bound of $r!e$. We do not know what the largest number is such that all integers up to this can be partitioned into s sets, no set containing the difference of two of its elements.

11. We have already referred to sequences of integers containing infinitely many primes. To show the existence of arbitrarily many primes, we do not have to investigate too many numbers. We will show that however we choose distinct positive integers, if we take all pairwise sums of these numbers and consider their prime divisors, there are arbitrarily many of these, assuming that we start with enough integers. More precisely, we prove the following theorem:

---

[4] Even a rough survey of these topics would considerably surpass the framework of this book, and instead we refer the interested reader to S. SINGH: *Fermat's Enigma*, Walker & Co. New York, 1997.

Theorem 7 *However we choose $2^k + 1$ distinct positive integers, if we take all pairwise sums, then at least $k + 1$ primes occur as prime divisors of these numbers.*

The main obstacle to finding many distinct prime divisors of these pairwise sums is the possibility that a prime to a high power can divide a sum even if it does not divide some of the summands. For this reason we try to find, for a prime p, many pairwise sums that are not divisible by higher powers of p than that of the summands. It is then expected that the pairwise sums will have many distinct prime divisors. In order to reach our goal, we will first prove a lemma.

Let an odd number of positive integers

$$n_1, n_2, \ldots, n_{2\ell+1}$$

be given, and let p be an odd prime. Write the numbers in the form

$$n_i = p^{\alpha_i} q_i \quad (i = 1, 2, \ldots, 2\ell + 1),$$

where the q_i's are not divisible by p. Then the sum of two of them, say n_i and n_j assuming $\alpha_i \leq \alpha_j$, is

$$n_i + n_j = p^{\alpha_i}(q_i + p^{\alpha_j - \alpha_i} q_j).$$

If $\alpha_i \neq \alpha_j$, then the sum in parentheses is not divisible by p. If, however, $\alpha_i = \alpha_j$, then we need to take care that $q_i + q_j$ is not divisible by p. This would be guaranteed if either both were smaller than $p/2$, or if both summands were greater than $p/2$ (since p is odd, they cannot equal $p/2$). In these cases, the sum of the two has remainder between 0 and p, respectively between p and $2p$. The endpoints are not possible, and the sum therefore is not divisible by p. The numbers

$$q_1, q_2, \ldots, q_{2\ell+1},$$

upon division by p must have either $\ell + 1$ remainders smaller than $p/2$, or $\ell + 1$ remainders greater than $p/2$. With this we have proved the following lemma:

Lemma *If p is an odd prime, then among any $2\ell + 1$ distinct odd integers there are $\ell + 1$ such that the pairwise sums of the numbers are not divisible by a higher power of p than the summands themselves.*

12. This lemma is the key to proving the theorem. Let $2^k + 1$ distinct positive integers be given (where k is at least 1). Among the prime divisors of all pairwise sums generated by these integers appears the number 2, since we have at least 3 numbers, and at least two of them are even or at least two of them odd, and the sum of these is even. Let us denote the odd prime divisors of the pairwise sums by $p_1, p_2, \ldots, p_i$. We will show that $i \geq k$ by indirect reasoning.

Assume that $i < k$. Now apply the above lemma. According to the lemma, from the $2^k + 1$ numbers we can choose $2^{k-1} + 1$ such that if the sum of any two of them is divisible by p_1^α, then both would be divisible by it. From these we can choose $2^{k-2} + 1$ numbers from them such that in addition to the above condition, if the sum of any two of them is divisible by p_2^β, then each of the summands is divisible by it.

Repeatedly applying this procedure we get $2^{k-i} + 1$ numbers, and by our assumption this means that at least 3 numbers such that the properties stated above on p_1 and p_2 hold for all $p_1, p_2, \ldots, p_i$. Let 3 of the remaining numbers be a_1, a_2 and a_3. Since we assumed that we have listed all the odd prime divisors for all pairwise sums, we have

$$a_1 + a_2 = 2^{u_0} p_1^{u_1} p_2^{u_2} \cdots p_i^{u_i}, \quad a_1 + a_3 = 2^{v_0} p_1^{v_1} p_2^{v_2} \cdots p_i^{v_i},$$
$$a_2 + a_3 = 2^{w_0} p_1^{w_1} p_2^{w_2} \cdots p_i^{w_i}.$$

Here both a_1 and a_2 have to be divisible by $p_1^{u_1} p_2^{u_2} \cdots p_i^{u_i}$, and hence neither of them can also be divisible by 2^{u_0}. Their sum can be divisible only by higher powers of 2 than those of the summands if 2 has the same exponent in both of their canonical representations. The same holds for a_1 and a_3. Therefore, the three numbers are of the following form:

$$a_i = 2^t b_i, \quad i = 1, 2, 3,$$

where b_1, b_2 and b_3 are odd; their pairwise sums have the form

$$b_1 + b_2 = 2^{r_1} p_1^{u_1} p_2^{u_2} \cdots p_i^{u_i},$$
$$b_1 + b_3 = 2^{r_2} p_1^{v_1} p_2^{v_2} \cdots p_i^{v_i}$$
$$b_2 + b_3 = 2^{r_3} p_1^{w_1} p_2^{w_2} \cdots p_i^{w_i}$$

(with $r_1 = u_0 - t$, $r_2 = v_0 - t$, and $r_3 = w_0 - t$). For the b's, it also holds that, for instance, b_1 and b_2 are divisible by $p_1^{u_1} p_2^{u_2} \cdots p_i^{u_i}$; hence none of them is less than this value. In fact, at least one of them is greater, since b_1 and b_2 are distinct.

Thus, r_1 needs to be at least 2. Similarly, both r_2 and r_3 are greater than or equal to 2. Therefore,

$$b_1 + b_2 = 4x, \quad b_1 + b_3 = 4y, \quad b_2 + b_3 = 4z,$$

for some x, y, and z. Dividing the sum of these by 2, we have

$$b_1 + b_2 + b_3 = 2x + 2y + 2z.$$

This is a contradiction, since on the left-hand side we have the sum of three odd numbers, which is odd, while on the right we have the sum of three even numbers, which is even. Therefore, we cannot have $i < k$. This proves Theorem 7.

This theorem was discovered by ERDŐS and TURÁN back when they were university students.[5] The limit of $2^k + 1$ seems as though it could be significantly decreased, but the theorem has not yet been proved for a smaller value.

It is natural to ask the question, given two sequences $1 < a_1 < a_2 < \cdots < a_k$ and $1 < b_1 < b_2 < \cdots < b_\ell$, what can we say about the number of prime divisors of $a_i + b_j$? GYŐRY, STEWART, and TIJDEMAN proved that if $k \geq \ell \geq 2$, then there is a positive constant c_1 such that the number of these prime divisors is greater than[6]

$$c_1 \log k.$$

On the other hand, ERDŐS, STEWART, and TIJDEMAN proved that this constant cannot be replaced by

$$c_2 (\log k)^2 \log \log k,$$

for any c_2 greater than $\frac{1}{8}$.[7]

The corresponding problem for products has also been investigated. In this case, the number of prime divisors of $a_i b_j + 1$ is being estimated. For this, GYŐRY, SÁRKÖZY, and STEWART have shown that (using the above notation) there is a similar bound of the form

$$c_3 \log k$$

for an appropriate positive constant c_3.[8]

13. We have already referred to sequences whose elements are pairwise relatively prime. In Section 5.4 we mentioned that the sequence of prime numbers is essentially the densest such sequence. Let us now weaken the requirement and investigate those sequences for which no element is a multiple of any other. Call such sequences *multiple-free*, or for short *M-sequences*. If the elements are pairwise relatively prime, then the new requirement is met, yet we can multiply every element of a pairwise relatively prime sequence by a fixed number m, and the new sequence is still an M-sequence, but its elements are not pairwise relatively prime anymore, since the greatest common divisor of every two elements is m.

[5] P. ERDŐS, P. TURÁN: *American Math. Monthly* **41** (1934), pp. 608–611. There they have $3 \cdot 2^{k-1}$ for our $2^k + 1$.

[6] K. GYŐRY, C. L. STEWART, R. TIJDEMAN: *Compositio Math.* **59** (1986), pp. 81–89.

[7] P. ERDŐS, C. L. STEWART, R. TIJDEMAN: *Compositio Math.* **66** (1988), pp. 1–71.

[8] K. GYŐRY, A. SÁRKÖZY, C. L. STEWART: *Acta Arithmetica* **74** (1996), pp. 365–385.

Denote by $M(N)$ the maximal number of elements in an M-sequence whose elements are all less than or equal to N. If $N = 2k$ or $N = 2k + 1$, then the numbers $k + 1, k + 2, \ldots, N$ form an M-sequence. In this way, we choose k numbers up to $2k$, and $k+1$ numbers up to $2k+1$, which is $[(N+1)/2]$ in both cases. Hence we have demonstrated that

$$M(N) \geq \left[\frac{N+1}{2}\right]. \tag{6}$$

We will show that equality holds here.

Theorem 8 *Let N be a positive integer. If none of the elements of a sequence consisting of integers not greater than N is a divisor of some other element, then the sequence has at most $[(N+1)/2]$ elements.*

14. We give two proofs to the above theorem. First we prove it by induction on N. It is clear that $M(1) = M(2) = 1$. Then it is enough to show that

$$M(N + 2) \leq M(N) + 1, \tag{7}$$

since this will imply that the inequality

$$M(N) \leq \left[\frac{N+1}{2}\right] \tag{8}$$

holds for all integers N, which in conjunction with (6) yields the desired equality. Suppose that (8) holds for two consecutive values of N, say $R - 1$ and R. Then by (7) and (8), we have

$$M(R + 1) \leq M(R - 1) + 1 \leq \left[\frac{R}{2}\right] + 1 = \left[\frac{R}{2} + 1\right] = \left[\frac{R+2}{2}\right],$$

and similarly

$$M(R + 2) \leq M(R) + 1 \leq \left[\frac{R+1}{2}\right] + 1 = \left[\frac{R+1}{2} + 1\right] = \left[\frac{R+3}{2}\right].$$

15. For another proof, we start from an M-sequence of numbers not larger than $N + 2$. If at most one of $N + 1$ and $N + 2$ belongs to the sequence, then the other numbers are not larger than N, and their number is at most $M(N)$. In this case (7) is true.

If $N+1$ and $N+2$ both appear in the sequence, then at least one is even; call this $2n$. Then neither n nor its divisors can appear in the sequence. This means that the elements not larger than N, along with the number n, also form an M-sequence. Thus there are at most $M(N) - 1$ elements not greater than N and therefore at most $M(N) + 1$ in the entire sequence. With this we have established (7) and proved Theorem 8.

16. We can rephrase the claim of Theorem 8 as follows: No matter how we choose $[(N + 1)/2]$ positive numbers between 1 and N, there are always two such that one is a divisor of the other. We will prove this statement using the earlier idea of writing a number as product of a power of two and an odd integer. Let the given numbers be

$$a_1, a_2, \ldots, a_m.$$

Here $m > [(N + 1)/2]$. Write these numbers as

$$a_i = 2^{u_i} b_i \quad (1 \leq i \leq m),$$

where u_i is a nonnegative integer and b_i an odd integer.

The number of odd integers not larger than N is $[(N + 1)/2]$. Thus there are two b's that are equal, say $b_i = b_j (i \neq j)$. We may assume that $u_i > u_j$, since $a_i \neq a_j$. Thus

$$a_i = 2^{u_i - u_j} 2^{u_j} b_j = 2^{u_i - u_j} a_j,$$

and so $a_j \mid a_i$. With this we have proved the claim.

17. The first proof given above is due to P. TURÁN and D. LÁZÁR, who found it independently of each other; the second proof is due to E. VÁZSONYI and M. WACHSBERGER.[9]

It can be shown that if an M-sequence of elements not larger than N has $[(N + 1)/2]$ elements, then every element is at least 2^v, where v is the integer satisfying

$$3^v \leq N < 3^{v+1}.$$

This means that as N grows, the smallest element of the sequence grows beyond all bounds. Thus there cannot be an infinite M-sequence for which the elements not larger than N form a maximal M-sequence. (In the case of sequences with pairwise relatively prime elements, the sequence of primes is such a sequence; in the case of sequences not containing the difference of any two elements, the sequence of odd integers is also a sequence with this property.)

Various problems arise in considering infinite M-sequences. What can be said of the density of such a sequence, if an infinite M-sequence even exists? Such sequences do exist; in fact, the sequence of primes is such a sequence. We know that this sequence has 0 density. Is it possible to give a denser infinite M-sequence, for example one that has positive upper density? Is there one whose lower density is also positive?

It is clear that if $a_1, a_2, \ldots$ is an infinite M-sequence, and we denote by $A(x)$ the number of elements not larger than x, then

---

[9] Compare these with P. TURÁN: *Középisk. Mat. Lapok*, new series **8** (1954), pp. 33–41 (in Hungarian).

$$A(x) \le M(x) = \left[\frac{x+1}{2} \right] \le \frac{x+1}{2},$$

so that

$$\frac{A(x)}{x} \le \frac{1}{2} + \frac{1}{2x}.$$

Thus for any arbitrarily small positive h, this is smaller than $\frac{1}{2} + h$ for x large enough, showing that the upper density is at most $\frac{1}{2}$. At the beginning of this section we mentioned that from a large x on, it is never possible that $A(x) = M(x)$. BEHREND and ERDŐS, each using different methods, proved that this equality is not even close to the truth; they showed that every infinite M-sequence has lower density 0,[10] and that the upper density is strictly less than $\frac{1}{2}$. In the other direction, BESICOVITCH exhibited an M-sequence for any arbitrarily small positive h that has upper density at least $\frac{1}{2} - h$.[11]

Exercise:

   11. Does there exist an M-sequence such that any two elements have a common divisor greater than 1, yet there is no number greater than 1 that divides all numbers greater then any given bound?

18. The sequences discussed in Section 6 do not contain the difference of any two elements of the sequence. A similar question, that just recently came to a comfortable resting point after tedious examinations, is the search for sequences that do not contain arithmetic progressions of length 3, or more generally, for a given k, the sequences that do not contain arithmetic progressions of length k. We will denote by $r_k(n)$ the maximal number of elements of sequences whose elements are not greater than n, and that do not contain arithmetic progressions of length k. This function behaves very whimsically.

Let us try to determine the first few values of $r_3(n)$. To make things shorter, call such a sequence a T-sequence. The following property of T-sequences makes our examination a bit easier. If

$$0 < a_1 < a_2 < \cdots < a_r$$

is a T-sequence of integers not greater than N, and $k < a_1$, then the following two sequences satisfy the same conditions:

(a) $a_1 - k, \ a_2 - k, \ \ldots, \ a_r - k,$

(b) $N + 1 - a_r, \ N + 1 - a_{r-1}, \ \ldots, \ N + 1 - a_1.$

---

[10] F. BEHREND: *Journal London Math. Soc.* **10** (1935), pp. 42–44. P. ERDŐS: *ibid.* **10** (1935) pp. 126–128. In both references, upper bounds are given to sums of the elements, from which it follows that the lower density is 0.

[11] A. S. BESICOVITCH: *Math. Annalen* **110** (1934), pp. 336–341.

It is clear that these contain only positive integers not greater than N, and the two equalities

$$(a_u - k) + (a_v - k) = 2(a_w - k)$$

and

$$(N + 1 - a_u) + (N + 1 - a_v) = 2(N + 1 - a_w)$$

hold if and only if

$$a_u + a_v = 2a_w.$$

Therefore, if either of the sequences (a) and (b) contains an arithmetic progression of length 3, then the original sequence would contain one as well, contradicting our assumption.

By the above property (a), we can assume that the smallest element of the sequence is 1.

19. Let us now determine the first few values of $r_3(N)$. It is obvious that

$$r_3(1) = 1, \quad r_3(2) = r_3(3) = 2, \quad r_3(4) = 3,$$

since 1, 2, and 3 cannot all belong to a T-sequence, but 1, 2, and 4 form a T-sequence.

We will show that

$$r_3(5) = r_3(6) = r_3(7) = r_3(8) = 4.$$

On the one hand, the sequence 1, 2, 4, 5 is a T-sequence with elements not greater than 5, and hence

$$r_3(8) \geq r_3(7) \geq r_3(6) \geq r_3(5) \geq 4.$$

On the other hand, if we show that from positive integers not greater than 8 it is not possible to form a T-sequence of length 5, then we have proved the above equality.

Such a sequence of length 5 would either have three elements not greater than 4, or three elements greater than 4. It is enough to consider the first case, because if the second case holds, then subtracting these elements from 9, by property (b) we would get a T-sequence of length 5 with elements not greater than 8 for which the first case holds.

The only T-sequences of length 3 containing elements not greater than 4 are 1, 2, 4 and 1, 3, 4. We cannot add 6 to the first sequence, because then 2, 4, and 6 forms an arithmetic progression. Similarly, we cannot add 7 to any of these sequences, because then 1, 4, and 7 forms an arithmetic progression. From 5 and 8 we can choose at most one to add to the previous sequences, because of the arithmetic progression 2, 5, 8.

To the sequence 1, 3, 4 we cannot add 5, because 3, 4, 5 is an arithmetic progression, nor 7, because of the 1, 4, 7; and only one of 6 or 8 can be added to 1, 3, 4, because of 4, 6, 8. So indeed, it is not possible to pick 5 integers between 1 and 8 that form a T-sequence, and thus

$$r_3(8) = 4 = r_3(7) = r_3(6) = r_3(5).$$

Here the question arises, first considered by I. SCHUR, for given k and ℓ, if we partition the positive integers up to a large enough bound into ℓ sets, is it true that one set contains k-term arithmetic progressions? VAN DER WAERDEN answered this question in the affirmative in 1928, giving an elementary, though rather deep, proof.[12] The following is a simple and interesting application of the theorem:

Exercise:

12. Prove that if $a_1, a_2, \ldots$ is an infinite sequence whose elements are either 1 or -1, then for every positive K, there exist numbers b, c, and d for which

$$\left| \sum_{i=k}^{d} a_{kb+c} \right| > K.$$

20. The values of the function $r_3(n)$ calculated thus far are all greater than $n/2$. For larger n this inequality will turn around, but to show this, it is necessary to determine the values of $r_3(n)$ up to $n = 22$, which is tiring work. With further work, it is possible to decrease the ratio from $\frac{1}{2}$ to $\frac{3}{8}$. ERDŐS and TURÁN conjectured more than 60 years ago that the ratio $r_k(n)/n$ will be smaller than any positive h, for all n larger than some constant depending on h and k. The path we have been exploring does not give any promise of leading to a proof of this stronger statement even for the case $k = 3$, since we have not discovered any pattern at all.

The first major result in this direction was by ROTH, who verified the conjecture for the case $k = 3$, and in fact proved the stronger statement that if n is large enough, then there is a constant c such that $r_3(n) < cn/\log\log n$. His method cannot be applied to larger values of k. SZEMERÉDI was the first to prove the conjecture for $k = 4$, and then finally in 1974 for all k.[13] The theorem was later proved using an entirely different method (using ergodic theory) by FURSTENBERG,[14] and the proof was later considerably shortened together with KATZNELSON and ORNSTEIN.[15]

Tighter bounds, similar to that achieved for r_3, are not known for $k > 3$ even though it is possible that $r_k(n)$ divided by $n/(\log n)^t$ is not bounded

[12] B. L. VAN DER WAERDEN: *Nieuw Arch. Wisk.* **15** (1927), pp. 212–216.
[13] E. SZEMERÉDI: *Acta Arithmetica* **27** (1975), pp. 199–245. The earlier literature is referred to here.
[14] H. FURSTENBERG: *Journ. Analyse Mat.*, **31** (1977), pp. 204–256.
[15] H. FURSTENBERG, Y. KATZNELSON, D. S. ORNSTEIN: *Proc. Symp. Pure Math.* **(39)** 1980, pp. 217–242.

no matter how large t is. If this were true, it would imply, for example, that there are arbitrarily long arithmetic progressions among the primes. Such an arithmetic progression of length 21, found in 1990, is the following:

$$142\,072\,321\,123 + 1\,413\,769\,024\,680 \cdot k, \qquad k = 0, 1, \ldots, 20.$$

21. Another sequence that has been considerably studied is the so-called *Sidon sequence*. These are sequences of nonnegative integers with the property that all pairwise sums of elements are distinct. These first arose during S. SIDON's investigation of Fourier sequences.[16] For short, let us call these sequences *S*-sequences.

For instance, the powers of 2 form an *S*-sequence, but one can give denser *S*-sequences than this one. Denote by $s(n)$ the maximum number of elements of an *S*-sequence with elements not greater than n, and let

$$1 \le a_1 < a_2 < \cdots < a_s \le n$$

be an *S*-sequence with $s = s(n)$ elements. Then all the sums $a_i + a_j$ are distinct, not greater than $2n$, and their number is $\binom{s+1}{2}$ (we do not exclude the case where $i = j$). Thus

$$\binom{s+1}{2} \le 2n,$$

from which it follows that

$$s < 2\sqrt{n}.$$

We get a better bound by observing that a sequence is an *S*-sequence if and only if the differences between any two elements are distinct. We leave the proof of this statement to the reader. In a fashion similar to that above, we get

$$\binom{s}{2} < n,$$

which implies that

$$s < \sqrt{2n + 1}.$$

22. Using even cleverer ideas, we can get rid of the factor $\sqrt{2}$ in the above bound. We will prove the following theorem in two different ways.

---

[16] S. SIDON: *Math. Annalen* **106** (1932), pp. 536–539. For results on *S*-sequences see P. ERDŐS, A. SÁRKÖZY, and V. T. SÓS: *Journ. of Number Theory* **47** (1994), pp. 329–347.

Theorem 9 *The maximal number of elements of an S-sequence with elements not greater than n satisfies*

$$s(n) < \sqrt{n} + \sqrt[4]{n} + 1.$$

Both proofs are based on estimating a sum related to s in two different ways. Again, we use the property of an S-sequence that the differences of all pairs of elements are distinct.

For an appropriate t to be determined later, consider the $n + t$ intervals of length $t - 1$ that intersect the interval $[0, n]$:

$$[-t + 1, 0], \ [-t + 2, 1], \ \ldots, \ [n, n + t - 1].$$

Let the number of elements of our S-sequence that fall into each of these intervals be $A_1, A_2, \ldots, A_{n+t}$. Each a_i falls into t consecutive intervals, so

$$\sum_{i=1}^{n+t} A_i = ts.$$

Now let us count the number of times the pair (a_i, a_j) (for $i > j$) falls within the above listed intervals. Let the total number of these be D. Then, on the one hand, it is clear that

$$D = \sum_{i=1}^{n+t} \binom{A_i}{2} = \frac{1}{2}\sum_{i=1}^{n+t} A_i^2 - \frac{1}{2}\sum_{i=1}^{n+t} A_i.$$

On the other hand, if the difference of a pair of elements is d, then this pair falls within $t - d$ intervals. Since all the differences are distinct, then each d can occur at most once. Therefore,

$$D \le \sum_{d=1}^{t-1}(t - d) = \frac{t(t-1)}{2}.$$

Comparing the above two relations for D, we have

$$\sum_{i=1}^{n+t} A_i^2 - \sum_{i=1}^{n+t} A_i \le t(t-1).$$

We saw that the second sum on the left-hand side equals ts. Now we apply the inequality for arithmetic and quadratic means to the first sum on the left-hand side:

$$\sum_{i=1}^{n+t} A_i^2 \ge \frac{\left(\sum_{i=1}^{n+t} A_i\right)^2}{n+t} = \frac{t^2 s^2}{n+t}.$$

Writing these into the above inequality, reducing it to zero, and multiplying both sides by $(n+t)/t^2$ we get that

$$s^2 - s\left(\frac{n}{t}+1\right) - \left(\frac{n}{t}+1\right)(t-1) \le 0.$$

For the values of s satisfying this second-degree inequality we have

$$s \le \frac{n}{2t} + \frac{1}{2} + \sqrt{n+t+\frac{n^2}{4t^2} - \frac{n}{2t} - \frac{3}{4}}.$$

Now, if we choose $t = \left\lceil \sqrt[4]{n^3} \right\rceil + 1$, then the first term on the right-hand side is less than $\frac{1}{2}\sqrt[4]{n}$, while the expression under the square root is less than the square of $\sqrt{n} + \frac{1}{2}\sqrt[4]{n} + \frac{1}{2}$. This yields the desired inequality.

23. For the second proof of the theorem, again using a parameter t to be determined later, we estimate the sum

$$K = \sum_{1\le i-j\le t} (a_i - a_j)$$

in two different ways. If $1 \le \mu \le t$, then a value μ can appear among the $i-j$ values in $s-\mu$ different ways. Hence the number of terms in the sum above is

$$(s-1) + (s-2) + \cdots + (s-t) = ts - \frac{t(t-1)}{2} = tw,$$

where

$$w = s - \frac{t+1}{2}.$$

All the differences are distinct; hence on the one hand their sum is not smaller than the sum of the first tw integers:

$$K \ge \frac{tw(tw+1)}{2} > \frac{t^2w^2}{2}.$$

On the other hand, those differences for which the difference of the indices is μ can be rearranged into so-called telescopic sums,

$$(a_{s-\sigma} - a_{s-\sigma-\mu}) + (a_{s-\sigma-\mu} - a_{s-\sigma-2\mu}) + \cdots < a_{s-\sigma} \le n.$$

Here we have that

$$0 \le \sigma < \mu,$$

so we have μ such telescopic sums, and naturally, we continue all of them as long as $s - \sigma - r\mu$ remains nonnegative. In this way, every term of K belongs to one of the sums. Since $1 \le \mu \le t$, this yields

$$1 + 2 + \cdots + t = \frac{t(t+1)}{2}$$

telescopic partial sums, and therefore

$$K < \frac{nt(t+1)}{2}.$$

Combining the two inequalities for K and multiplying both sides by $2/t^2$ we arrive at

$$w^2 = \left(s - \frac{t+1}{2}\right)^2 < n + \frac{n}{t}.$$

Solving for s, we have

$$s < \frac{t+1}{2} + \sqrt{n + \frac{n}{t}},$$

and by choosing

$$t = \left[\sqrt[4]{n}\right] + 1$$

we arrive at the bound that we wanted to prove.

The first proof is due to Erdős and Turán,[17] while the second is due to Lindström.[18]

24. Theorem 9 is sharp in the sense that the ratio $s(n)/\sqrt{n}$ is arbitrarily close to 1 for n large enough. In other words, the asymptotic value of $s(n)$ is $\sqrt{n}$. To prove this last statement, we will prove the following lower bound on $s(n)$:

Theorem 10 *There exists an S-sequence with s elements not greater than n for which*

$$s > \sqrt{n} - n^{11/40}.$$

There are several known proofs of this fact. Here, we present the proof requiring the least background. Every known proof is related to the one we present, and is based on the following theorem interesting on its own:

Theorem 11 *Let p be an odd prime. There exist $p-1$ numbers a_i for which the differences $a_i - a_j$ $(i \neq j)$ are incongruent modulo $p^2 - p$.*

Let g be a primitive root, modulo p (see Theorem 2.18), and let the a_i's be the smallest nonnegative solution, modulo $p^2 - p$, to the simultaneous congruences

$$x \equiv i \pmod{p-1},$$

$$x \equiv g^i \pmod{p}.$$

We need to show that the congruence $a_i - a_j \equiv a_r - a_s \pmod{p^2 - p}$, or written in the equivalent form

---

[17] P. Erdős, P. Turán: *Journ. London Math. Soc.* **16** (1941), pp. 212–215.
[18] B. Lindström: *Journ. Comb. Theory* **6** (1969), pp. 211–212.

$$a_i + a_s \equiv a_r + a_j \pmod{p^2 - p},$$

is satisfied only by the trivial solutions. In other words, this means that for any number c there is at most one pair of numbers i, j that satisfies the congruence

$$c \equiv a_i + a_j \pmod{p^2 - p}.$$

Based on the definition of the a_i's, this is equivalent to the congruences

$$c \equiv i + j \pmod{p - 1},$$

$$c \equiv g^i + g^j \pmod{p}$$

being simultaneously satisfied. The first congruence we may rewrite as

$$g^c \equiv g^i g^j \pmod{p}.$$

The relationship between the roots of a quadratic equation and its coefficients implies that the residue classes $(g^i)_p$ and $(g^j)_p$ are uniquely defined as the two roots of the second-degree congruence

$$x^2 - cx + g^c \equiv 0 \pmod{p}.$$

Since the modulus is prime, the pair of roots is uniquely determined, and thus the pair i, j is uniquely defined.

25. The sequence of a_i's constructed is an S-sequence, and for n of the form $p^2 - p$, we see that

$$s(n) \leq p - 1 \leq \frac{1}{2}\left(\sqrt{4n+1} + 1\right) - 1 > \sqrt{n} - 1.$$

For arbitrary n, we choose a prime such that $p^2 - p$ is close to n. This essentially means that p is close to $\sqrt{n}$. Chebyshev's theorem, which states that there is a prime between m and $2m$, for every m, gives too large of a gap. We mentioned the conjecture that for every positive δ there is a prime between m and $m + m^\delta$, but the proof of this looks to be without hope. Here we will use the deep result that for every m sufficiently large there is a prime between $m - m^{11/20}$ and m.[19] Thus we can choose a prime p between $\sqrt{n} - n^{11/40}$ and $\sqrt{n}$, and hence

$$s(n) \geq s\left(p^2 - p\right) \geq p - 1 \geq \sqrt{n} - n^{11/40}.$$

Exercises:

13. Prove that there is only one 8-term T-sequence of positive integers not greater than 14.

[19] R. C. Baker, G. Harman: *Proc. London Math. Soc.* (3) **72** (1996), pp. 261–280.

14. Denote by $t(N)$ the maximum number of elements not greater than N in a sequence not containing arithmetic progressions of length 5, and for which every arithmetic progression of length 3 is part of an arithmetic progression of length 4. Determine the values of t for N not greater than 14.

15*. Let $0 < a_1 < a_2 < \cdots$ be an infinite sequence and assume that there is a K such that for every n,

$$\sum_{i=1}^{n} \frac{1}{a_i} < K.$$

Prove that a sequence of numbers not divisible by any of the a_i's has asymptotic density.

16. The first two elements of a sequence of integers are 1 and 2, and the sequence does not contain the sum of any two of its elements. Prove that this sequence has at most $(n + 5)/3$ elements not greater than n (O. Lukács).

17. Show that there are infinite sequences of integers $1 < a_1 < a_2 < \cdots$ such that no element is the divisor of the product of the previous elements, and for which
$$\pi(a_k) \le k.$$

Are there sequences for which the inequality is strict except for finitely many k?

18*. Take the sequence of integers that do not have any prime divisors other than those from a finite set of previously given primes. Prove that this sequence does not have any infinite subsequence in which no element is a divisor of any other (Dickson).

19. Prove that from any infinite sequence of positive integers we can either choose an infinite subsequence in which every element is the divisor of the following one, or one in which no element is the divisor of any other.

20*. Let $a_1, a_2, \ldots, a_{k\ell+1}$ be a sequence of distinct real numbers. Prove that there is either an increasing subsequence of $k + 1$ elements or a decreasing subsequence of $\ell + 1$ elements (Erdős–Szekeres).

21*. Prove that among any $k\ell + 1$ distinct integers we may either choose $k+1$ such that each is divisible by the previous one, or we may choose $\ell + 1$ such that none is the divisor of any other.

22. Prove that it is possible to give n positive integers $a_1, a_2, \ldots, a_n$, but not more, not greater than $2n$ such that no integer between 1 and $2n$ is divisible by two different a_i's.

23. Consider all numbers of the form $a_0 + 3a_1 + 9a_2 + \cdots + 3^k a_k$, where each a_i is either 0 or 1. Prove that there are no three numbers in this sequence that form an arithmetic progression (Szekeres).

7. Diophantine Problems

1. We have already encountered Diophantine equations in Chapter 1. For example, we investigated in detail first-degree equations with two unknowns (Sections 1.29–1.30), and found a presentation for all Pythagorean triples (Sections 1.21–1.24). In Chapter 4 we studied which positive integers can be represented as the sum of two squares. We will return to the question in Section 4. We will now consider a collection of similar questions.

One analogue of right triangles in space is the right parallelepiped. The lengths of the edges, x_1, x_2, x_3, and the diagonal, y, satisfy the equation

$$x_1^2 + x_2^2 + x_3^2 = y^2.$$

The integer solutions can be called "Pythagorean quadruples." In what follows, we will find for any integer n (greater than 1) the integer solutions in parametric form of the more general equation

$$x_1^2 + x_2^2 + \cdots + x_n^2 = y^2. \tag{1}$$

It is clearly enough to determine positive solutions ($x_i > 0, y > 0$) for which $(x_1, x_2, \ldots, y) = 1$. Let us rewrite the equation as

$$x_1^2 + x_2^2 + \cdots + x_{n-1}^2 = (y - x_n)(y + x_n), \tag{2}$$

and let d be the greatest common divisor $(x_1, x_2, \ldots, x_{n-1}, y - x_n)$. Then for appropriate integers $u_1, u_2, \ldots, u_n$ for which $(u_1, u_2, \ldots, u_n) = 1$, we have

$$x_i = du_i, \qquad i = 1, 2, \ldots, n - 1; \qquad y - x_n = du_n.$$

Making these substitutions and simplifying by d, we may write (2) as

$$d(u_1^2 + u_2^2 + \cdots + u_{n-1}^2) = u_n(du_n + 2x_n) = u_n(2y - du_n).$$

After rearranging a little, we have

$$2u_n x_n = d(u_1^2 + u_2^2 + \cdots + u_{n-1}^2 - u_n^2), \tag{3}$$

$$2u_n y = d(u_1^2 + u_2^2 + \cdots + u_{n-1}^2 + u_n^2). \tag{4}$$

Here $(x_n, d) = 1$, since any common divisor is a common divisor of x_1, $x_2, \ldots, x_{n-1}$ as well, and we assumed that $(x_1, x_2, \ldots, x_n) = 1$. Thus x_n is a divisor of the multiplier of d on the right-hand side of (3). In other words, for an appropriate integer v, we have

$$u_1^2 + \cdots + u_{n-1}^2 - u_n^2 = vx_n \quad \text{and} \quad 2u_n = dv.$$

Writing these into our earlier equations, we have

$$vx_i = dvu_i = 2u_iu_n \quad (i = 1, 2, \ldots, n-1), \tag{5_i}$$

$$vx_n = u_1^2 + \cdots + u_{n-1}^2 - u_n^2. \tag{5_n}$$

Multiplying (4) by v we obtain

$$2vu_ny = dv\left(u_1^2 + \cdots + u_n^2\right) = 2u_n\left(u_1^2 + \cdots + u_n^2\right),$$

which we may simplify by dividing through by $2u_n$ to obtain

$$vy = u_1^2 + \cdots + u_n^2. \tag{$5'$}$$

Here we must choose v so that $(x_1, \ldots, x_n) = 1$, which means that

$$v = \left(2u_1u_n, \ldots, 2u_{n-1}u_n, u_1^2 + \cdots + u_{n-1}^2 - u_n^2\right). \tag{6}$$

It is easy to convince ourselves that the $x_1, \ldots, x_n, y$ obtained for any real values of the u_i's satisfy (1).

Since $x_n > 0$, it is enough to find those systems of parameters for which

$$u_n^2 < u_1^2 + \cdots + u_{n-1}^2. \tag{7}$$

We note as well that it is not necessary to require that the u_i be relatively prime, since otherwise, only v is multiplied by the square of their possible common divisor.

2. We will finally show that a given solution set can arise from only one series of ratios $u_1 : u_2 : \cdots : u_n$. Since we are interested only in the ratios, we may assume that

$$x_i = u_i, \quad i = 1, 2, \ldots, n-1,$$

and we will find all appropriate u_n's.

We have

$$vx_i = 2u_nx_i, \quad (i = 1, 2, \ldots, n-1)$$

and hence

$$v = 2u_n.$$

Thus

$$2u_nx_n = x_1^2 + \cdots + x_{n-1}^2 - u_n^2 = y^2 - x_n^2 - u_n^2,$$

from which it follows that

$$0 = y^2 - x_n^2 - 2x_n u_n - u_n^2 = (y - x_n - u_n)(y + x_n + u_n).$$

Thus for u_n, we have arrived at two possible values:

$$u_n = y - x_n \qquad \text{or} \qquad u_n = -y - x_n.$$

Only the first of these is positive, which determines the value of u_n. We have shown the following:

Theorem 1 *All integer solutions to* (1) *are given by the formulas* $(5_1), \ldots,$ $(5_n), (5')$, *with the conditions given in* (6) *and* (7).

A system of integer solutions can arise from only one system of parameters, up to a proportion factor.

It would be enough for x_n to write the absolute value on the right-hand side. With this, (7) becomes superfluous but one solution will then be provided by two different systems of parameters. The second arises if we start with the parameter $d = (x_1, \ldots, x_{n-1}, y + x_n)$.[1]

Exercises:

1. What procedure arises from this proof for the representation of the Pythagorean triples?

2. The solution obtained here clearly depends on the order of the parameters. For $n = 3$ and $n = 4$, give infinitely many examples of parameters that yield the same solution regardless of their order.

3*. Determine a parametric form for the integer solutions of the equation

$$x^2 + y^2 = z^2 + u^2.$$

3. Our results concern representations of square numbers as sums of squares. In Chapter 4 we briefly discussed when a number can be represented as the sum of two squares. It was easy to see that numbers of the form $4k - 1$ are not the sum of two squares (Theorem 4.13). There are also numbers that are not the sum of three squares. For example, $7 = 4 + 1 + 1 + 1$ is the decomposition of 7 as the sum of fewest squares.

This observation gives rise to many questions. Which integers can be written as the sum of three squares? Is every integer the sum of 4, 5, or 6 squares? Is there a number g such that every positive integer can be written as the sum of at most g squares?

---

[1] The parametric forms differs slightly from the formula for Pythagorean triples in that there only the four arithmetic operations occur, while here the proportion factor v can be determined as a greatest common multiple and the parameters u_i are unique only up to a proportion factor.

In what follows we will answer these questions, but new similar questions arise. For example, which numbers are representable as the sum of k positive squares $(k = 1, 2, \ldots)$? Is there a number h such that every integer large enough can be represented as the sum of h positive squares?

4. Based on previous theorems, we will give a complete solution to the following question: When does the equation

$$x^2 + y^2 = n \tag{8}$$

have an integral solution? We will use the following identity first observed by EULER:

$$\left(x^2 + y^2\right)\left(u^2 + v^2\right) = (xu - yv)^2 + (xv + yu)^2. \tag{9}$$

Applied to our problem, this identity says that if two numbers can both be represented as the sum of two squares, then so can their product. By simply repeating this argument, we see that the product of arbitrarily many numbers, each of which is the sum of two squares, is also the sum of two squares.

It now follows from Theorem 4.13, using the fact that $2 = 1^2 + 1^2$, that every number that is not divisible by a prime of the form $4k - 1$ is the sum of two squares. If we multiply such a number by an arbitrary square, then observing that the product of squares is itself a square, we see that the new number is still the sum of two squares. Thus the product can be divisible by primes of the form $4k - 1$, but always to an even power. With this we have seen that (8) has an integer solution if all primes of the form $4k - 1$ in the canonical decomposition of n appear to an even power.

5. EULER was the first to show that these conditions are also necessary, i.e., that only numbers of this form can be written as the sum of two squares.

Theorem 2 *All numbers for which primes of the form $4k - 1$ do not appear to an odd power in their canonical decomposition, and only these numbers, can be represented as the sum of two squares.*

Let a and b be two positive integers, and let c be their largest common divisor:

$$a = ca_1, \quad b = cb_1, \quad \text{where} \quad (a_1, b_1) = 1.$$

The sum of their squares is

$$a^2 + b^2 = c^2 \left(a_1^2 + b_1^2\right).$$

The sum of squares on the right-hand side cannot be divisible by a prime p of the form $4k - 1$, for otherwise, according to Theorem 1.16, it would not be relatively prime to one of the bases, and since p is prime, this would mean that it divides one of the bases. But then p could divide the sum only if it also divides the other term, which would mean that p divides the other base. This implies that p is a common divisor of a_1 and b_1, which we assumed to be relatively prime. Thus the power of p that divides $a^2 + b^2$ is the same power that divides c^2, and is hence even. With this we have proved Theorem 2.

6. The condition in Theorem 2 is a very strong restriction. There are infinitely many numbers for which the condition does not hold, for example the primes of the form $4k - 1$, but there are many others. Checking small numbers, we see that these "bad numbers" occur rather frequently. It can be proved that this is the case for larger numbers as well. How does the situation improve if we allow three squares instead of two? For 7, three is still not enough, but it is easy to find many exceptions as well.

In the case of the sum of two squares, we were able to obtain a great deal of information from looking at the residues of the squares modulo 4. For the case of three squares, it is useful to look at the residues modulo 8. The square of an even number is divisible by 4; hence either it is divisible by 8 (has a remainder of 0) or it gives a remainder of 4. The squares of odd numbers, on the other hand, always give a remainder of 1 modulo 8 (Theorem 1.14).

Thus, the sum of three numbers all even, respectively two even, respectively one even, respectively all odd, have a residue (modulo 8) of 0 or 4, respectively 1 or 5, respectively 2 or 6, respectively 3. We see that the sum of three squares can never have a residue of 7 modulo 8. The numbers that have remainder 7 upon division by 8 are therefore not the sum of three squares. In fact, a little more than this is true.

Theorem 3 *Numbers of the form $4^n(8k+7)$, where n and k are nonnegative integers, cannot be represented as the sum of the squares of three integers.*

Let us assume that a number $N = 4^n(8k + 7)$ were the sum of three squares, and let n be the smallest exponent for which this is true. We have already shown that n cannot be 0. We have also seen that the sum of three squares is even only if the three numbers are all even, or if one is even and the other two are odd. In the latter case, the sum is not divisible by 4.

Thus if the sum is divisible by 4, then all three bases are even. Taking half of each base, we get $N' = 4^{n-1}(8k + 7)$ as the sum of three squares, which contradicts the assumption that n is the smallest exponent for which this is possible. With this we have proved Theorem 3.

7. Are all numbers not of the form given in the theorem the sum of three squares? This question was answered positively by GAUSS, but his proof used very refined arguments and is beyond the scope of this book.[2]

---

[2] See, for example, E. LANDAU: *Vorlesungen über Zahlentheorie*, I.S. Hirzel, Leipzig, 1927, pp. 114–125.

Every number is the sum of four squares. It was LAGRANGE who first proved this fact. Since than, many different proofs have been given. We will present one here.

Theorem 4 *Every positive integer can be represented as the sum of four squares.*

As in the proof of Theorem 3, that of Theorem 4 will reduce to the case of primes, using the following identity, also due to EULER:[3]

$$(x^2 + y^2 + z^2 + u^2)(r^2 + s^2 + t^2 + v^2)$$
$$= (xr + ys + zt + uv)^2 + (xs - yr - zv + ut)^2 \qquad (10)$$
$$+ (xt + yv - zr - us)^2 + (xv - yt + zs - ur)^2.$$

It follows from this that if two numbers can each be written as the sum of four squares, then so can their product. Thus Theorem 4 will follow from a special case:

Theorem 4′ *Every prime can be represented as the sum of four squares.*

8. The previous theorem will follow easily using the two lemmas below, both of which relate to congruences.

Lemma 1 *For every prime p, the congruence*

$$x^2 + y^2 + 1 \equiv 0 \pmod{p} \qquad (11)$$

has a solution.

Lemma 2 *If m and r are integers greater than 1, and we are given r linear expressions with integer coefficients*

$$L_i(u_1, u_2, \ldots, u_k) = a_{i1}u_1 + a_{i2}u_2 + \cdots + a_{ik}u_k \qquad (12)$$

$$(i = 1, 2, \ldots, r),$$

where $k > r$, and further $h_1, h_2, \ldots, h_k$ are positive real numbers satisfying the inequality

$$h_1 h_2 \cdots h_k \geq m^r,$$

then the system of congruences

$$L_i \equiv 0 \pmod{m} \qquad (i = 1, 2, \ldots, r)$$

has an integer solution $u_1, u_2, \ldots, u_k$ for which

$$|u_i| \leq h_i \qquad (i = 1, 2, \ldots, k),$$

and not all the u_i are zero.

---

[3] Identity (9) can be seen as the special case of this where $z = u = t = v = 0$.

Let us first see how to prove Theorem 4' using these lemmas. Let p be an arbitrary prime. Because of the identities

$$2 = 1^2 + 1^2 \ (+0^2 + 0^2) \qquad \text{and} \qquad 3 = 1^2 + 1^2 + 1^2 \ (+0^2),$$

we will assume that $p > 3$. According to Lemma 1, there are integers x and y for which (11) is true.

We will apply Lemma 2 to the linear forms $L_1 = xu_1 + yu_2 - u_3$ and $L_2 = -yu_1 + xu_2 - u_4$ with $m = p$ and $h_i = \sqrt{p}$ ($i = 1, 2, 3, 4$). Then

$$h_1 h_2 h_3 h_4 = p^2,$$

and according to the lemma, there are integers u_1, u_2, u_3, u_4, not all 0, and (because $\sqrt{p}$ is not an integer)

$$|u_i| < \sqrt{p} \qquad (i = 1, 2, 3, 4).$$

Multiplying congruence (11) by $\left(u_1^2 + u_2^2\right)$ and applying the identity (9), we have

$$\left(u_1^2 + u_2^2\right)\left(x^2 + y^2\right) + u_1^2 + u_2^2 = (xu_1 + yu_2)^2 + (xu_2 - yu_1)^2 + u_1^2 + u_2^2$$
$$\equiv u_3^2 + u_4^2 + u_1^2 + u_2^2 \equiv 0 \pmod{p}.$$

Thus we have on the one hand that $u_1^2 + u_2^2 + u_3^2 + u_4^2$ is divisible by p, and on the other hand that

$$0 < u_1^2 + u_2^2 + u_3^2 + u_4^2 < 4p,$$

and hence the sum of the squares is $p, 2p$, or $3p$.

9. In the first case, we have the desired expression. In the second case, among the four bases, either at least two are even, or at least two are odd. Assume that u_1 and u_2 are of the same parity. Then this holds for u_3 and u_4 as well, for otherwise, the sum would be odd. Multiplying the sum of squares by 2, we have

$$4p = (u_1 + u_2)^2 + (u_1 - u_2)^2 + (u_3 + u_4)^2 + (u_3 - u_4)^2.$$

Every base is even, and dividing each by 2, we have the desired expression for p.

If finally the sum of the squares is equal to $3p$, then one of the bases is divisible by 3, and the other three are not. This follows from the fact that the square of an integer n not divisible by 3 has a remainder of 1 upon division by 3. This is immediate from the expression

$$n^2 - 1 = (n - 1)(n + 1)$$

and the observation that one of the terms on the right is divisible by 3, since one of every three consecutive integers is divisible by 3 and we assumed that n is not.

It is not possible that every base is divisible by 3, since this would mean that the sum of the squares is divisible by 9, but the prime p is greater than 3, and thus $3p$ is not divisible by 9.

If the number of bases not divisible by 3 were 1, 2, 3, or 4, then the sum of the squares would have remainder 1, 2, 0, and 1, respectively. Thus in fact, the sum of the squares is $3p$ only for the case mentioned.

We may assume that u_4 is divisible by 3 and that the other three have remainder 1 upon division by 3; if any of the terms had remainder 2, we could replace it by its negative. We may easily verify the following expression for three times the sum of the squares:[4]

$$9p = (u_1 + u_2 + u_3)^2 + (u_1 - u_2 + u_4)^2 + (u_2 - u_3 + u_4)^2 + (u_3 - u_1 + u_4)^2.$$

Here every term in parentheses is divisible by 3, and taking one-third of each base we obtain the desired expression for p.

10. In order to prove Theorem 4' and with it Theorem 4, we still need to prove the two lemmas.

To prove the first lemma, in the case of a prime p larger than 3, let us consider the following two sequences of $(p + 1)/2$ numbers:

$$1, 1^2 + 1, 2^2 + 1, \ldots, ((p-1)/2)^2 + 1; \qquad 0, -1^2, -2^2, \ldots, -((p-1)/2)^2.$$

Together there are $p+1$ numbers, and hence at least two must be in the same residue class modulo p.

No two numbers from the first sequence, nor from the second sequence, can be in the same residue class. This would mean that there exist distinct integers x_1 and x_2 for which

$$|x_i| \le \frac{p-1}{2}, \quad i = 1, 2,$$

and further

$$x_1^2 + 1 \equiv x_2^2 + 1 \pmod{p}, \qquad \text{or} \qquad -x_1^2 \equiv -x_2^2 \pmod{p}.$$

Both of these congruences are equivalent to

$$x_1^2 - x_2^2 = (x_1 - x_2)(x_1 + x_2) \equiv 0 \pmod{p}.$$

For a prime p, this is possible only if one of the two factors is congruent to 0. But this is not possible, because

---

[4] This is, in fact, just identity (10) in the case where one of the factors is $3 = 1^2 + 1^2 + 1^2 + 0^2$.

$$0 < |x_1 - x_2| < (p-1)/2 \quad \text{and} \quad 0 < x_1 + x_2 < p - 1.$$

Thus the two elements that are congruent are from different sequences. In other words, there are numbers x and y such that

$$x^2 + 1 \equiv -y^2 \pmod{p},$$

which is just congruence (11) rearranged.

This theorem and the proof given are due to EULER.

11. The proof of Lemma 2 will run similarly to that of Lemma 1, but rather than investigating the residues of sequences of numbers, we will investigate the residues of r-tuples of numbers. For integers m and r, both greater than 1, and the first-degree expressions (12), we consider the sequence of r-tuples of numbers

$$L_1, L_2, \ldots, L_r,$$

where

$$L_i = L_i(z_1, z_2, \ldots, z_k),$$

and for $i = 1, 2, \ldots, k$, every z_i takes all values $0, 1, \ldots, [h_i]$. According to the hypothesis, the number of such r-tuples satisfies

$$([h_1] + 1)([h_2] + 1) \cdots ([h_k] + 1) > h_1 h_2 \cdots h_k \geq m^r.$$

If we consider the remainders modulo m (for example, the smallest non-negative remainders) of the components of the r-tuples, there are m^r possible values. There are more r-tuples than this, and thus there must be two that have the same residues for all corresponding components, in other words they are congruent modulo m. Thus there are two r-tuples corresponding to parameters $z_1, z_2, \ldots, z_k$ and $z_1', z_2', \ldots, z_k'$. As a consequence of these observations, we have

$$a_{i1} z_1 + a_{i2} z_2 + \cdots + a_{ik} z_k \equiv a_{i1} z_1' + a_{i2} z_2' + \cdots + a_{ik} z_k' \pmod{m},$$

for $i = 1, 2, \ldots, r$, and thus for the u_j defined as

$$u_j = z_j - z_j', \quad j = 1, 2, \ldots, k,$$

it follows that

$$L_i(u_1, u_2, \ldots, u_k) \equiv 0 \pmod{m}, \quad i = 1, 2, \ldots, r.$$

Additionally, we have

$$|u_j| = |z_j - z_j'| \leq h_j, \quad j = 1, 2, \ldots, k,$$

and not every u_j is zero, since the two k-tuples of parameters are different. With this we have proved all the claims of the lemma.

A special case of this lemma was first given by THUE and used to study representations of integers as the sum of two squares.

Exercises:

We know from earlier results (Theorem 2.15) that for primes of the form $4k + 1$, there is a number a satisfying $a^2 + 1 \equiv 0 \pmod{p}$, and hence for these, Lemma 1 is true.

4. Prove Lemma 1 for primes of the form $4k - 1$ using the fact that -1 is not a quadratic residue.

5. Prove using Lemma 2 that primes of the form $4k+1$ can be decomposed as the sum of two squares.

6. Prove that the representation of primes of the form $4k + 1$ as the sum of two squares is unique, disregarding the order and the signs of the bases.

12. According to Theorem 4, every natural number can be decomposed into two summands each of which is the sum of at most two squares. From Theorem 2 it follows that in the canonical decompositions of the two summands, primes of the form $4k - 1$ do not occur to odd powers. Thus *every positive integer can be decomposed into two summands such that in their canonical decompositions, all primes of the form $4k - 1$ occur to even powers.* Conversely, if we could prove this statement independently, then Theorem 4 would follow from this and Theorem 2. EULER spent a long time trying unsuccessfully to prove Theorem 4 in this way, and to the best of our knowledge, such a proof has never been demonstrated.

In addition to the question of when a number can be written as the sum of two squares, the number of ways of writing a number in this way has also been studied. As an example, we mention without proof the following interesting result: *For a number n, let τ_1 and τ_3 be the numbers of divisors of the form $4k + 1$ and $4k + 3$, respectively. Then the number of ways of writing n as the sum of two squares is $4(\tau_1 - \tau_3)$.* In determining the number of ways of writing an integer as the sum of two squares, it was helpful that primes of the form $4k + 1$ essentially have a unique representation (see Exercise 6). In the case of four squares, this is no longer true. For example,

$$31 = 5^2 + 2^2 + 1^2 + 1^2 = 3^2 + 3^2 + 3^2 + 2^2,$$

$$37 = 6^2 + 1^2(+0^2 + 0^2) = 5^2 + 2^2 + 2^2 + 2^2 = 4^2 + 4^2 + 2^1 + 1^2,$$

$$67 = 8^2 + 1^2 + 1^2 + 1^2 = 7^2 + 4^2 + 1^2 + 1^2 = 7^2 + 3^2 + 3^2 \ (+0^2)$$

$$= 5^2 + 5^2 + 4^2 + 1^2,$$

$$103 = 10^2 + 1^2 + 1^2 + 1^2 = 9^2 + 3^3 + 3^2 + 2^2 = 7^2 + 7^2 + 2^2 + 1^2$$

$$= 7^2 + 6^2 + 3^2 + 3^2 = 7^2 + 5^2 + 5^2 + 2^2.$$

We will no longer consider the representation of integers as the sum of squares.[5] The more general question of representing integers as the sum of kth powers, for an arbitrary k, was first considered by WARING, who conjectured that for every k there is a number $n(k)$, such that every integer is representable by at most $n(k)$ kth powers.[6] HILBERT was the first to prove WARING's conjecture in 1909.

13. The question for third powers is quite difficult and goes beyond the scope of this book, yet at the same time, the question of representations by sums of fourth powers can be led back to representations by squares. (The difference in difficulty is similar to that between the sum of three squares and the sum of four squares.) We start our investigation with the following identity:

$$(a + b)^4 + (a - b)^4 = 2 \left(a^4 + 6a^2b^2 + b^4 \right).$$

On the right-hand side there are only second and fourth powers, and thus we may arrive at a perfect square if we can achieve, for example, that a^4 appears two more times, without a^2b^2 appearing again. Of course this also holds for b and for all other parameters to be introduced. We can reach our goal by introducing parameters c and d, and for all 6 pairs of the four, we combine the above identities:

$$(a + b)^4 + (a - b)^4 + (a + c)^4 + (a - c)^4 + (a + d)^4 + (a - d)^4$$
$$+ (b + c)^4 + (b - c)^4 + (b + d)^4 + (b - d)^4 + (c + d)^4 + (c - d)^4$$
$$= 6 \left(a^4 + b^4 + c^4 + d^4 + 2a^2b^2 + 2a^2c^2 + 2a^2d^2 + 2b^2c^2 + 2b^2d^2 + 2c^2d^2 \right)$$
$$= 6 \left(a^2 + b^2 + c^2 + d^2 \right)^2.$$

Applying Theorem 4, according to which every number can be written as the sum of four squares, the identity above shows that six times a square can be written as the sum of 12 fourth powers.

Referring again to Theorem 4, every positive integer can be written as the sum of four squares. Thus we can represent all multiples of 6 by at most 48 fourth powers.

Finally, dividing a number by 6 gives a remainder $r \leq 5$, and by adding that many 1's $(= 1^4)$, we can build up every natural number using at most 53 fourth powers. With this we have shown the following:

---

[5] For additional reading regarding the number of representations of integers as the sum of two squares, see, for example, I. NIVEN, H. S. ZUCKERMAN: *An Introduction to the Theory of Numbers*, Wiley, New York, 1991. A more general reference treating a broader range of questions is E. GROSSWALD: *Representations of Integers as Sums of Squares*, Springer, New York, 1985.

[6] E. WARING: *Meditationes Arithmeticae*, Cambridge, 1770, Theor. XLVII.

Theorem 5 *Every positive integer can be represented as the sum of at most 53 fourth powers.*

This theorem and essentially the proof we presented were found by J. L. LAGRANGE (1736–1813).

14. The number 53, as one would expect, is a great deal larger than the expected value. (Using a simple idea, we can immediately reduce it to 50. See Exercise 8.) On the other hand, we can show that there are infinitely many numbers in whose representation as the sum of fourth powers at least 16 fourth powers are necessary. Namely, as KEMPNER observed, the following holds:

Theorem 6 *Numbers of the form $16^n \cdot 31$ $(n = 0, 1, 2, \dots)$ cannot be represented as the sum of fewer than 16 fourth powers.*

The validity of this claim is immediately clear for the case $n = 0$, i.e., for 31. In its representation, only 1^4 and 2^4 can appear. The latter can appear at most once, and the remaining 15 can only be made by putting 1's together.

From here, we will prove the claim indirectly. For this purpose, let us see what the possible remainders are for fourth powers of integers, modulo 16. The fourth powers of even integers are divisible by 16, hence congruent to 0. For an odd number $2k + 1$, we have

$$(2k + 1)^4 = 16k^4 + 32k^3 + 24k^2 + 8k + 1 \equiv 8k(k + 1) + 1 \equiv 1 \pmod{16},$$

because one of k and $k + 1$ is even.

Let us now assume that there is an n for which a number of the form given in the theorem can be represented as the sum of at most 15 fourth powers, and let m be the smallest such exponent. We know that m cannot be 0, and hence the number is divisible by 16. Thus every summand must be even, since the remainder modulo 16 is the number of odd summands, and this number is smaller than 16. However, taking half of all the bases would give a representation of the number corresponding to $n = m - 1$ as the sum of at most 15 fourth powers, contradicting our assumption that m was the smallest exponent. Thus the theorem is true for all exponents.

15. The upper and lower bounds we have found for the number of terms are quite far from one another. By trying to represent many numbers, one can guess that the lower bound is closer to the minimum number of necessary terms. After 31, the representation of 47 again requires a "large" number of terms, it requires 17; 18 are needed for 63, and 79 cannot be represented as the sum of fewer than 19 fourth powers. Further trials do not lead to more unfavorable cases.

Let us denote by $g(k)$ the least upper bound on the number of kth powers necessary to represent positive integers. We have already seen that

$$g(2) = 4, \qquad 19 \leq g(4) \leq 53.$$

It is expected, and supported by experimentation, that the number $g(k)$ is determined by small numbers, up to which there are few kth powers, and hence we must use many of these to give the desired representation. Based on this idea, EULER considered the numbers whose representation cannot contain kth powers other than 1^k and 2^k, and obtained the lower bound

$$g(k) \geq 2^k + \left[\left(\frac{3}{2} \right)^k \right] - 2.$$

WIEFERICH showed in 1909 that equality holds here for $k = 3$, i.e., that $g(3) = 9$. A missing part of his proof was later supplied by KEMPNER in 1912.

Because $g(k)$ is essentially determined by small numbers, it is more characteristic to state Waring's problem as that of determining the number $G(k)$ of kth powers that are necessary to represent every large enough natural number, and such that there are infinitely many numbers that cannot be represented by fewer than $G(k)$ kth powers. (Theorem 3 shows that $G(2)$ is also 4, and according to Theorem 6, we have $G(4) \geq 16$.)

During the last century intensive research has gone on in the study of Waring's problem. Here we mention only some of the important results. HARDY and LITTLEWOOD, in some of their papers between 1919 and 1928, gave a series of results using complex functions that provided upper bounds for $G(k)$; their methods proved to be useful for treating other questions as well. Later, by combining this with a new method, VINOGRADOV showed how these results might be improved. Using these ideas, VAUGHAN gave the best bound on $G(k)$ to date:[7]

$$G(k) < k(3 \log k + 4.2).$$

For larger k, this bound is smaller that the lower bound given above for $g(k)$. This leads to the result that equality holds in the bound for $g(k)$ for $k \geq 7$ as long as some other condition is not satisfied. If this condition is satisfied, then $g(k)$ is given by a formula similar to that above. Another 40 years were needed to determine the value of $g(k)$ for the missing three cases: $k = 4, 5, 6$. The final one, that $g(4) = 19$, was proved in 1986 by BALASUBRAMANIAN, DESHUILLERS, and DRESS.[8] All that is known of the condition mentioned above is that only finitely many k can satisfy it and that the smallest such k is larger than $200\,000$; it seems rather likely that it is never satisfied.

HARDY and LITTLEWOOD conjectured that $G(k) \leq 2k + 1$, except for when $k = 2^m$ and $m \geq 2$, in which case they conjectured that $G(k) = 4k$.

---

[7] R. C. VAUGHAN: *Acta Math.* **162** (1989), pp. 1–71.

[8] R. BALASUBRAMANIAN, J-M. DESHUILLERS, F. DRESS: *Comptes Rendus, Acad. Sci. Paris*; Ser. 1 **303** (1986), pp. 85–88 and pp. 161–163.

The results to date do not contradict this conjecture, yet it does hold for $k = 2$ and 4, since we know that $G(4) = 16$.[9]

It is helpful in determining the value of $G(k)$ if we know how many numbers up to a bound x can be written as the sum of k terms of kth powers. LANDAU showed that for $k = 2$, as x grows beyond all bounds, this number is asymptotically $cx/\sqrt{\log x}$, where c is a positive constant (thus in any case the desired numbers have density 0). For $k \geq 3$, we do not even know whether the sequence of such numbers has density 0.

It can be conjectured that the number of positive integers that may be written as the sum of k terms of kth powers for a given k and an arbitrary positive h is greater than x^{1-h}, as long as x is large enough. DAVENPORT proved in 1950 that for $k = 3$, this number is larger than $x^{47/55-h}$ if x is large enough. ERDŐS and MAHLER showed that the number of positive integers of the form $a^k + b^k$ not greater than x is greater than $cx^{2/k}$ for a suitable constant c, and smaller than $x^{2/k}$; thus in this case the exact order of magnitude is known.[10]

HARDY and LITTLEWOOD conjectured that for any given h, the number of solutions of the equation

$$n = x_1^k + x_2^k + \cdots + x_k^k$$

is smaller than n^h, as long as n is large enough. This is true for $k = 2$, but for $k = 3$ MAHLER refuted it. We do not know what the situation is for $k > 3$. It is possible that the conjecture does not hold in these cases either.[11]

Exercises:

7. Prove that there are infinitely many numbers in addition to those in Theorem 6 that cannot be represented as the sum of fewer than 16 fourth powers.

8. Prove that every positive integer can be represented as the sum of at most 50 fourth powers.

9*. Prove that $G(k) \geq k$.

16. We will arrive at some problems, in some respect easier than the previous ones, by still investigating the representation of integers in terms of kth powers ($k \geq 2$), but now allowing subtraction as well as addition. We will restrict ourselves to the representations of nonnegative integers, since the

---

[9] G. H. HARDY, E. M. WRIGHT: *An Introduction to the Theory of Numbers*, Oxford (1960), Chapters XX and XXI; T. D. WOOLEY: *Annals of Math.* **135** (1992) pp. 131–164.

[10] P. ERDŐS, K. MAHLER: *Journal London Math. Soc.* **13** (1938), pp. 134–139.

[11] See, for example, J.-M. DESHOUILLERS: *Seminar on Number Theory*, 1984–1985. Exp. No. **14** 47 pp., Univ. Bordeaux I, Talance, 1985; N. D. ELKIES: *Math. Comp.*, **51** (1988), pp. 825–835.

representation of a negative integer can be obtained from the representation of its absolute value by reversing all the signs.

The representation of an integer by kth powers with the same sign is a special case of this mixed-sign representation, and hence the result of HILBERT mentioned earlier implies that for every positive integer k there exists a number W such that every positive integer can be represented by adding and subtracting at most W terms of kth powers. Denote the smallest such number of necessary terms by $w(k)$. It is clear that

$$w(k) \le g(k),$$

but it is expected that the value of $w(k)$ is considerably smaller than that of $g(k)$.

Let us first investigate the case of $k = 2$. The difference between two consecutive squares is

$$n^2 - (n-1)^2 = 2n - 1,$$

and hence every odd number can be written in this way. For even numbers,

$$2n = (2n - 1) + 1 = n^2 - (2n - 1)^2 + 1^2,$$

and hence at most 3 squares suffice to represent every integer:

$$w(g) \le 3.$$

It is easy to see that there are infinitely many integers that require 3 squares in this representation. We know that there are infinitely many positive integers that cannot be represented as the sum of two squares. Is it possible that these numbers are representable as the difference of two squares? The difference of two squares is

$$a^2 - b^2 = (a + b)(a - b),$$

and here, observing that $(a + b) + (a - b) = 2a$, we see that either both of the factors are even, or both are odd. The product is thus either divisible by 4, or it is odd. Twice an odd integer cannot be written as the product of two factors of the same parity, and hence neither can the difference of two squares. Such a number is 6, since it is divisible by 3, but not by 9. Thus we have seen that

$$w(2) \ge 3.$$

It is also true that there are infinitely many numbers for which three numbers are necessary in this type of representation, for example, numbers of the form $6(2k + 1)^2$, which are twice an odd integer and at the same time have an odd power of 3 in the canonical decomposition, and are hence neither the difference nor the sum of two squares. With this we have proven the following:

Theorem 7 *Every integer can be represented as a 3-term sum of squares and inverses of squares. In addition, there are infinitely many integers that are not representable as the sum of two squares.*

17. For third powers we can get a good bound with the help of the following identity:

$$(m + 1)^3 - 2m^3 + (m - 1)^3 = 6m.$$

This shows that all multiples of 6 are representable by at most 4 cubes.

From the identity

$$n^3 - n = (n + 1)n(n - 1)$$

we see that for integers n, the right-hand side is divisible by 6, because among the factors there is an even one and one divisible by 3, and 2 and 3 are relatively prime to each other. This difference is therefore representable by four cubes. With this we have a representation of any integer n in terms of 5 cubes. Thus we have

$$w(3) \leq 5.$$

We can get a lower bound by observing that the cube of a number not divisible by three is adjacent to a number divisible by 9, and of course the cube of a number divisible by 3 is divisible by 9. Indeed, a number not divisible by 3 can be written as $3t + e$, where e is either 1 or -1 (hence $e^2 = 1$ and $e^3 = e$). The cube of this is

$$(3t + e)^3 = 27t^3 + 27t^2 e + 9t + e = 9t \left(3t^2 + 3te + 1\right) + e,$$

which is adjacent to a number divisible by 9, as we claimed.

From this it follows that a number of the form $9m + 4$ cannot be represented by fewer than 4 cubes. Thus for cubic representations we have the following result:

Theorem 8 *Every integer can be represented as the sum (with mixed signs) of at most 5 cubes, and there are infinitely many numbers that cannot be written as the sum of fewer than 4 such terms.*

18. Rather easily we were able to show that the numbers $w(2)$ and $w(3)$ exist and to find bounds for them, in the first case determining the exact value. In a similar way one can show that $w(k)$ exists for all k, without having to refer to results for Waring's problem, based on the identity

$$\sum_{i=0}^{k-1} (-1)^{k-1-i} \binom{k - 1}{i} (n + i)^k = k!n + d,$$

where d is a constant depending only on k (see Exercise 10). The proof of the existence of $g(k)$, however, goes far beyond the scope of this book.

In contrast with the results we mentioned earlier for $g(k)$, the exact value of $w(k)$ is known only for $k = 2$. Beyond this, we know very little about this function.[12]

---

[12] G. H. HARDY and E. M. WRIGHT: *An Introduction to the Theory of Numbers*, Oxford (1960), §§XXI.7–XXI.8.

Exercises:

10. Prove the identity above, and using it, prove that

$$w(k) \le 2^{k-1} + \frac{1}{2}k!.$$

11. Accepting that $G(k)$ exists, prove that

$$w(k) \le G(k) + 1.$$

19. In connection with congruences, we have proved Wilson's theorem (Theorem 2.14). This said that if p is a prime number, then $(p-1)! + 1$ is divisible by p, and we know (see Exercise 2.22) that if m is a composite number larger than 4, then $m \mid (m-1)!$.

If p is one of the primes $2, 3, 5$, then the value of $(p-1)! + 1$ is $2, 3, 25$, respectively, a power of the prime in question. Does this happen for larger primes as well, i.e., are there primes p greater than 5 such that

$$(p-1)! + 1 = p^k \tag{13}$$

for some integer k?

We will show that this is not possible.

Theorem 9 *There is no prime p greater than 5 for which $(p-1)! + 1$ is a power of p.*

Subtracting 1 from each side of (13) and dividing by $p-1$, we have

$$(p-2)! = p^{k-1} + p^{k-2} + \cdots + p + 1.$$

According to our remarks, the left-hand side of the equation is divisible by $p-1$, because $p-1$ is an even number greater than 4, and hence composite.

Subtracting one from each summand on the right-hand side, each term becomes divisible by $p-1$; hence the remainder of the right-hand side upon division by $p-1$ is equal to the number of terms k; hence this number must be divisible by $p-1$, and is thus at least $p-1$. But already in (13), all the $p-1$ factors are smaller than p, and even with the summand 1, the left-hand side is much smaller than the right. Thus the equality cannot hold.

20. If instead of asking whether $(p-1)! + 1$ is a power of p, but rather asking whether it is divisible by p raised to a power greater than 1, then the following example has been known for a long time:

$$13^2 \mid 12! + 1.$$

Besides this example, of the primes less than $100\,000$, only 563 satisfies this condition.[13] It is not known whether there exist other primes with this property, yet it has not been excluded that there may be infinitely many.

A question concerning powers of integers that received much attention and remained unsolved for long time was the question of whether the product of consecutive integers could be a power of an integer. Recently, the problem has been mostly solved. An easier question, and the one we will consider here, is whether binomial coefficients can be perfect powers.

Theorem 10 *If k and n are integers and $n/2 \geq k \geq 4$, then $\binom{n}{k}$ is not an integer raised to a power greater than 1.*

We note that $\binom{k}{2}$ is a square for infinitely many values of k. The equations $x^2 = 2y^2 + 1$ and $x^2 = 2y^2 - 1$ have infinitely many solutions (see Exercise 16), and for these x and y, we have

$$\binom{x^2}{2} = (xy)^2, \quad \text{respectively} \quad \binom{2y^2 + 1}{2} = (xy)^2.$$

The only solution of the equation $\binom{u}{3} = x^2$, other than the trivial ones $u = 3, x = 1$ and $u = 4, x = 2$, is

$$\binom{50}{3} = \frac{50 \cdot 49 \cdot 48}{1 \cdot 2 \cdot 3} = 25 \cdot 49 \cdot 16 = 140^2.$$

The proof of this fact was given recently—following many older investigations that proved incomplete—by K. Győry using deep investigations. The proof that this is the only solution uses elementary methods, but is quite involved.[14]

The theorem holds for powers greater than 2 in the case of $k = 2$ and $k = 3$ as well. R. Oblàth showed that the theorem holds for third, fourth, and fifth powers.[15]

21*. We will prove Theorem 10 in an indirect way. In the proof we will use the theorem of Sylvester and Schur (Theorem 5.9′), which we stated without proof.

Let us assume that for some integers n, k, r, for which $n/2 \geq k \geq 4, r > 1$, there is an integer a such that

$$\binom{n}{k} = a^r.$$

For the binomial coefficients we have $\binom{n}{k} = \binom{n}{n-k}$, and here either k or $n - k$ is at most $n/2$. Thus we may assume that $n \geq 2k$.

---

[13] This was noted by I. Z. Ruzsa.

[14] K. Győry: *Acta Arithmetica* **80** (1997) pp. 285–295.

[15] R. Oblàth: *Journ. London Math. Soc.* **23** (1948), pp. 252–253.

From every factor in the numerator of

$$\binom{n}{k} = \frac{n(n-1)\cdots(n-k+1)}{k!},$$

let us factor out the largest rth power that divides it:

$$n - i = b_i c_i^r \quad (i = 0, 1, \ldots, k-1).$$

Here b_i is not divisible by an rth power greater than 1. We will see that b_i cannot have a prime divisor greater than k. If there is a prime p such that $p \mid a$ and $p > k$, then the denominator is not divisible by p, and in the numerator there are only k factors, and hence p can divide only one of the terms $n - i$. Thus if the binomial coefficient is the rth power of an integer, then in the canonical decomposition of the factor in question, p^r must appear to some power. This means that it is a factor of c_i^r, and b_i is not divisible by p, as we claimed.

The theorem of Sylvester and Schur referred to above states that the binomial coefficient has a prime divisor p greater than k. If this is a divisor of $n - i_0$, then by our observations, we have

$$n - i_0 = b_{i_0} c_{i_0}^r \geq p^r > k^r,$$

and hence

$$n \geq n - i_0 > k^r. \tag{14}$$

22*. We will show that the numbers b_i are all different, and their product is thus $k!$. If there were two b's that were equal, $b_i = b_j; 0 \leq i < j < k$, then on the one hand,

$$b_i c_i^r - b_j c_j^r = n - i - (n - j) = j - i < k,$$

and on the other hand, this difference would be positive and

$$b_i c_i^r - b_j c_j^r = b_j(c_i^r - c_j^r) \geq b_i((c_j + 1)^r - c_j^r) > b_j r c_j^{r-1}$$

$$\geq 2b_j c_j^{r-1} \geq 2b_j \sqrt{c_j^r} \geq 2\sqrt{b_j c_j^r} \geq 2\sqrt{n - k + 1}.$$

Comparing the two bounds we have

$$2\sqrt{n - k + 1} < k,$$

and because both sides are positive,

$$4(n+1) = 4(n - k + 1) + 4k < k^2 + 4k < (k+2)^2 < 4k^2, \quad n < k^2 - 1.$$

This contradicts (14), and hence we cannot have $b_i = b_j$ for $i \neq j$.

It immediately follows from this that

$$b_0 b_1 \cdots b_{k-1} \geq k!.$$

23*. We will show that equality must hold here because $k!$ is divisible by the left-hand side. To do this, we will again use the indirect assumption that the binomial coefficient is a power greater than 1 of a number a greater than 1.

The quotient of the two sides of the inequality above gives

$$\frac{b_0 b_1 \cdots b_{k-1}}{k!} = \frac{1}{(c_0 c_1 \cdots c_{k_0})^r} \binom{n}{k} = \frac{a^r}{(c_0 c_1 \cdots c_{k_0})^r} = \frac{u^r}{v^r},$$

where the fraction on the right is in reduced form; i.e., $(u, v) = 1$.

We saw that the factors of the numerator in the fraction on the far left are not divisible by primes greater than k, and this clearly holds for the denominator as well. Additionally, the factors of the numerator are not divisible by numbers—and hence not by primes—raised to a power greater than $r - 1$.

Let q be a prime not greater than k, whose power in the canonical decomposition of the product $b_0 b_1 \cdots b_{k-1}$ is s, and in that of $k!$ is s_1. We can determine these numbers by adding to the number of the k factors $n - i$ that are divisible by q the number divisible by q^2, by $q^3, \ldots$, and finally by q^{r-1}.

Among consecutive integers, those divisible by m occur at intervals of distance m, thus among the k numbers there are at most $[k/m] + 1$ divisible by m. Using this we see that

$$s \le \sum_{t=1}^{r-1} \left(\left[\frac{k}{q^t} \right] + 1 \right) = r - 1 + \sum_{t=1}^{r-1} \left[\frac{k}{q^t} \right].$$

Theorem 1.12, which related to factorials, leads to

$$s_1 = \sum_{t=1}^{\infty} \left[\frac{k}{q^t} \right] \ge \sum_{t=1}^{r-1} \left[\frac{k}{q^t} \right].$$

Thus the numerator of the fraction $(u/v)^r$ can be divisible by q to a power not greater than $r - 1$. The numerator is the rth power of an integer, and since q is a prime, it must be divisible by at least q^r if it is divisible by q. This shows that u^r is divisible neither by primes at most k, nor by primes greater than k, i.e., $u^r = 1, u = 1$, and thus

$$k! = b_0 b_1 \cdots b_{k-1} v^r \ge b_0 b_1 \cdots b_{k-1}.$$

This together with the inequality at the end of the previous section shows that

$$b_0 b_1 \cdots b_{k-1} = k!.$$

If $r = 2$ and $k \ge 4$, then we immediately arrive at a contradiction, since $4 = 2^2 (= 2^r)$ cannot occur among the b's, and since all b's are distinct, it must be the case that

$$b_0 b_1 \cdots b_{k-1} < k!,$$

contradicting the equality we just obtained.[16]

---

[16] In the same way we may finish the proof of the theorem for any r, if we restrict ourselves to those k such that $k \ge 2^r$.

24*. If $k \geq 4$ and $r \geq 3$, then among the b's, the numbers 1, 2, and 4 must occur. Thus there are indices u, v, w for which

$$n - u = c_u^r, \qquad n - v = 2c_v^r, \qquad n - w = 4c_w^r.$$

We will show that the difference $(n-u)(n-w) - (n-v)^2$ is not 0, and by estimating this difference we will arrive at a contradiction. Let us introduce the quantities m, x, y defined by

$$n - v = m, \qquad n - u = m + x, \qquad n - w = m + y.$$

Here $1 \leq |x| < k, 1 \leq |y| < k$. Substituting, we have

$$(n - u)(n - w) - (n - v)^2 = (m + x)(m + y) - m^2 = m(x + y) + xy.$$

If $x + y = 0$, then the right-hand side is not 0. If rather $x + y \neq 0$, then using (14), we have

$$|m(x + y)| \geq m > n - k > k^r - k > (k - 1)^2 > |xy|,$$

and in this case as well the quantity under consideration cannot be zero.

We will estimate the absolute value of the difference from both sides. For an upper bound, we have

$$|(n - v)^2 - (n - u)(n - w)| < n^2 - (n - k + 1)^2 < 2n(k - 1).$$

For a lower bound, let d be the smaller of c_v^2 and $c_u c_w$. Because the two values are not equal, the larger of the two must be at least $d + 1$. Additionally, the value of either $(2c_v^r)^2$ or $c_u^r \cdot 4c_w^r$ is $4d^r$, and hence it is always true that

$$4d^r \geq (n - k + 1)^2. \tag{15}$$

Using these, it follows that

$$|(n - v)^2 - (n - u)(n - w)| = 4|(c_v^{2r} - c_u c_w)^r| \geq 4((d + 1)^r - d^r) > 4rd^{r-1}.$$

Comparing the upper and lower bounds, multiplying by d, and using (15), we have

$$2n(k - 1)d > 4rd^2 \geq r(n - k + 1)^2 > r(n^2 - 2kn). \tag{16}$$

It is advantageous to write n^2 multiplied by a small constant in place of the last subtrahend. We need to estimate the subtrahend from above. Since $r \geq 3, k \geq 4$, and $n > k^r$, we have

$$kn = n^2 \frac{k}{n} < n^2 \left(\frac{1}{k}\right)^{r-1} \leq \frac{n^2}{k^2} \leq \frac{n^2}{16},$$

from which it follows that

$$2kn < \frac{1}{8}n^2,$$

and thus

$$r(n^2 - 2kn) > 3\left(n^2 - \frac{1}{8}n^2\right) = 3 \cdot \frac{7}{8}n^2 > 2n^2.$$

In order to simplify the form of the left-hand side of (16), we need to compare n and d, which we will accomplish with the help of comparing n and c_i:

$$c_i \le (n - i)^{1/r} \le n^{1/3},$$

from which it follows for any indices i, j, k that

$$d \le n^{2/3},$$

and hence

$$2n(k - 1)d < 2kn^{5/3}.$$

Thus the inequality (16) leads to the inequality

$$2n^2 < 2kn^{5/3},$$

i.e.,

$$n < k^3.$$

In the case $r \ge 3$, this contradicts inequality (14) as well, and hence the binomial coefficient $\binom{n}{k}$ cannot be an rth power for $r > 2$ whenever $n/2 \ge k \ge 4$, verifying Theorem 10.

25. In the beginning of Section 20 we mentioned that the question of whether consecutive integers could be powers of integers was an open problem for a long time. It is easy to see that if $n > 1$ and $k > 1$, then $n!$ is not a kth power. It is also easy to see that the product of two, three, or four consecutive integers cannot be a kth power of an integer (see Exercises 17 and 18). SZEKERES and SIVASANKARANARAYANA PILLAI proved this for the product of up to 8 consecutive integers. The proof is based on the following lemma: If $k \le 16$, then among any k consecutive integers there is always one that is relatively prime to all the others (see Exercise 19). A. BAUER and S. PILLAI showed that this is not true for any k greater than 16.

For squares, ERDŐS and RIGGE proved independently from each other in 1938 the general theorem. They also proved (again, independently from each other) that for every t, there cannot be infinitely many pairs (n, k) for which

$$n(n + 1) \cdots (n + k - 1) = x^t. \tag{17}$$

SIEGEL and ERDŐS also proved that there is a constant c such that (17) has no solutions for $k > c$. Finally, ERDŐS and SELFRIDGE were able to show[17] that (17) is not solvable for any $t > 1$.

---

[17] P. ERDŐS, J. L. SELFRIDGE: *Illinois Journ. of Math.* **19** (1975), pp. 292–301.

An obvious generalization of the question is whether the product of consecutive terms of an arithmetic progression can be a perfect power, and in the other direction, whether there are arithmetic progressions all of whose terms are rth powers. This last question has the solution $1, 25, 49$ for squares ($r = 2$), and more generally, $v^2, 25v^2, 49v^2, v = 1, 2, \ldots$, gives infinitely many 3-term arithmetic progressions of squares. There are also infinitely many relatively prime 3-term arithmetic progressions of squares. It is not difficult to give a parametric representation for all of these (see Exercise 24).

From the observations mentioned above, it follows that there are infinitely many arithmetic progressions in which there are three consecutive terms whose product is a perfect square. EULER proved, on the other hand, that there are no 4-term arithmetic progressions of squares. It can also be shown that the product of four consecutive terms of an arithmetic progression cannot be a square. A related conjecture of ERDŐS is that for every positive number h, an n-term arithmetic progression can contain at most hn squares for arbitrary n large enough.

The following question is also related to consecutive integers. Other than $8 = 2^3, 9 = 3^2$, do there exist consecutive integers that are both powers (greater than 1) of integers? Is it possible to have three consecutive integers that are all perfect powers? For the first question, which seemed inaccessible for a very long time, TIJDEMAN recently gave a nearly complete solution showing that $a^x - b^y = 1$ can be solved for at most finitely many integers a, b, x, and y.[18] It is easy to see, however, that four consecutive integers cannot be perfect powers (see Exercise 20).

Exercises:

12*. Prove that for every n, there is a number between n and $n + 4\sqrt[4]{n}$ that can be written as the sum of two squares.

Note. Here instead of 4 we may write, for example, $2\sqrt{2}$. It is expected that for an arbitrarily small positive number h, there is a number that can be written as the sum of two squares between n and hn, for n large enough. This problem is still undecided.

13. For a number m, we know that there are four consecutive integers such that the sum of the squares of the first two is still a square, and that the sum of the squares of the latter two is divisible by m. Determine all possible values of m.

14. Give infinitely many numbers, different from those in Theorem 6, that cannot be written as the sum of fewer than 16 fourth powers.

15. Prove that every integer can be written in infinitely many different ways in the following form:

---

[18] R. TIJDEMAN: *Acta Arith.* **29** (1976), pp. 197–209

$$\sum_{i=1}^{k} e_i i^2 \qquad (e_i = \pm 1, \quad i = 1, 2, \ldots, k).$$

16. Consider the sequences of numbers $x_{k+1} = 3x_k + 2y_k$, $y_{k+1} = 4x_k + 3y_k$, with starting values $x_0 = 0$, $y_0 = 1$, and $x_0 = y_0 = 1$, respectively. Show that for every n, in the first case the equality $2x_n^2 + 1 = y_n^2$ holds, and in the second case the equality $2x_n^2 - 1 = y_n^2$.

17*. Prove that the product of three consecutive positive integers cannot be a perfect power.

18*. Prove that the product of four consecutive positive integers cannot be:

(a) a perfect square,

(b) a perfect kth power, for any $k > 2$.

19*. Prove that among 16 consecutive integers there is always one that is relatively prime to the product of the others (Szekeres).

20. Prove that there are no four consecutive integers each of which is a power greater than 1 of an integer.

21. If $f(x)$ is a polynomial, let $\Delta f(x)$ be the polynomial $f(x+1) - f(x)$, and if $k = 1, 2, \ldots$, let $\Delta^{k+1} f(x) = \Delta(\Delta^k f(x))$.

(a) Prove that if f is a polynomial of degree n, then $\Delta^{n+1} f(x)$ is the 0 polynomial.

(b) Determine the polynomial $\Delta^n f(x)$.

22*. Let f be a polynomial of degree n with integer coefficients, assume that the coefficients of f are pairwise relatively prime, and that d is a divisor of the value of f for every integer. Prove that $d \mid n!$.

Note. The statement remains true if instead of "pairwise relatively prime" we just write "relatively prime."

23*. Denote by $a_1, b_1, c_1, a_2, b_2, c_2$ real numbers such that if x and y are integers, then at least one of the two linear forms

$$a_1 x + b_1 y + c_1 \qquad \text{and} \qquad a_2 x + b_2 y + c_2$$

is an even integer. Prove that all three of the coefficients of at least one of the two forms are integers (Gy. Hajós).

24. Give a parametric representation for all 3-term arithmetic progressions of perfect squares.

25*. Prove that the equation $x^4 + y^4 = z^2$ has no solutions in positive integers (Fermat).

Note. From this it follows that $x^4 + y^4 = z^4$ has no positive integer solutions. This is the special case of Fermat's last theorem when the exponent is 4.

26*. For the equation $x^y = y^x$, determine all
    (a) integer solutions,
    (b) rational solutions.

8. Arithmetic Functions

1. Earlier, we encountered the function $\tau(n)$, which gives the number of (positive) divisors of the number n, and we also met Euler's function $\varphi(n)$. Given the canonical decomposition of n we gave formulas to compute these functions. We also learned some interesting properties of Euler's function. Although the function $\pi(x)$, which gives the number of primes not exceeding x, is defined for all positive values x, its value can change only at integer values of the argument. Therefore, $\pi(x)$ can also be viewed as a function whose domain consists of the positive integers.

Functions defined on the domain of positive integers are called *arithmetic functions*. In what follows we shall mostly consider functions taking integer values, but in general we do not restrict the range. For the study of certain questions it is useful to consider complex-valued arithmetic functions. Of primary interest are the functions expressing some arithmetic property of the integers.

2. We based the calculation of Euler's function on the observation that this function is *multiplicative* (in the sense of Section 2.39, Theorem 2.19). We shall now give a procedure to calculate $\varphi(n)$ without using the multiplicative property. This will of course imply the multiplicative property.

Let the canonical decomposition of n be

$$n = p_1^{k_1} p_2^{k_2} \cdots p_r^{k_r},$$

where we assume that none of the exponents is zero. We shall repeatedly use the fact that the number of positive integers not greater than n and divisible by a number d is $[n/d]$.

Let $n_{i_1 i_2 \ldots i_s \bar{j}_1 \bar{j}_2 \ldots \bar{j}_t}$ denote the number of those integers not exceeding n that are divisible by $p_{i_1}, p_{i_2}, \ldots, p_{i_s}$ (and perhaps by other prime divisors of n) but not divisible by $p_{j_1}, p_{j_2}, \ldots, p_{j_t}$. Here $i_1, i_2, \ldots, i_s, j_1, j_2, \ldots, j_t$ are distinct integers not greater than r. We allow s or t to be 0; in this case, the corresponding subscripts do not occur. In particular, for $s = t = 0$ we obtain n. With this notation we have

$$\varphi(n) = n_{\bar{1}\bar{2}\ldots\bar{r}}.$$

It is obvious on the one hand that

$$n_{i_1 i_2 \ldots i_s j_1} + n_{i_1 i_2 \ldots i_s \bar{j}_1} = n_{i_1 i_2 \ldots i_s},$$

and on the other hand,

$$n_{i_1 i_2 \ldots i_s} = \frac{n}{p_{i_1} p_{i_2} \cdots p_{i_s}}.$$

From these two equations,

$$n_{i_1 i_2 \ldots i_s \bar{j}_1} = \frac{n}{p_{i_1} p_{i_2} \cdots p_{i_s}} - \frac{n}{p_{i_1} p_{i_2} \cdots p_{i_s} p_{j_1}} = \frac{n}{p_{i_1} p_{i_2} \cdots p_{i_s}} \left(1 - \frac{1}{p_{j_1}} \right).$$

Combining this result with the equation

$$n_{i_1 i_2 \ldots i_s \bar{j}_1 j_2} + n_{i_1 i_2 \ldots i_s \bar{j}_1 \bar{j}_2} = n_{i_1 i_2 \ldots i_s \bar{j}_1}$$

we obtain

$$
\begin{aligned}
n_{i_1 i_2 \ldots i_s \bar{j}_1 \bar{j}_2} &= \frac{n}{p_{i_1} p_{i_2} \cdots p_{i_s}} \left(1 - \frac{1}{p_{j_1}} \right) - \frac{n}{p_{i_1} p_{i_2} \cdots p_{i_s} p_{j_2}} \left(1 - \frac{1}{p_{j_1}} \right) \\
&= \frac{n}{p_{i_1} p_{i_2} \cdots p_{i_s}} \left(1 - \frac{1}{p_{j_1}} \right) \left(1 - \frac{1}{p_{j_2}} \right).
\end{aligned}
$$

Continuing this way we obtain for any t that

$$n_{i_1 i_2 \ldots i_s \bar{j}_1 \bar{j}_2 \ldots \bar{j}_t} = \frac{n}{p_{i_1} p_{i_2} \cdots p_{i_s}} \left(1 - \frac{1}{p_{j_1}} \right) \left(1 - \frac{1}{p_{j_2}} \right) \cdots \left(1 - \frac{1}{p_{j_t}} \right).$$

Hence we have solved our problem, since in particular, we have

$$
\begin{aligned}
\varphi(n) = n_{\bar{1}\bar{2}\ldots\bar{r}} &= n \left(1 - \frac{1}{p_1} \right) \left(1 - \frac{1}{p_2} \right) \cdots \left(1 - \frac{1}{p_r} \right) \\
&= p_1^{k_1 - 1} p_2^{k_2 - 1} \cdots p_r^{k_r - 1} (p_1 - 1)(p_2 - 1) \cdots (p_r - 1).
\end{aligned}
$$

3. The proof we discussed is in essence a probabilistic argument, since

$$v_{i_1 i_2 \ldots i_s \bar{j}_1 \bar{j}_2 \ldots \bar{j}_t} = \frac{n_{i_1 i_2 \ldots i_s \bar{j}_1 \bar{j}_2 \ldots \bar{j}_t}}{n}$$

is the probability that in choosing an integer at random from $1, 2 \ldots, n$, this integer will be divisible by the primes $p_{i_1}, p_{i_2}, \ldots, p_{i_s}$ and not divisible by the primes $p_{j_1}, p_{j_2}, \ldots, p_{j_t}$. Since the primes $p_1, p_2, \ldots, p_r$ are distinct and divide n, we have

$$v_{i_1 i_2 \ldots i_s} = \frac{1}{p_{i_1} p_{i_2} \cdots p_{i_s}} = v_{i_1} v_{i_2} \cdots v_{i_s}.$$

Let us call E_i the "event" that the number selected is divisible by p_i. Then the above equation expresses the fact that $E_1, \ldots E_r$ are independent. (We can say the same if instead of primes, we consider a family of pairwise relatively prime divisors of n.) Let $\bar{E}_i$ denote the event that E_i fails to occur. The probability of this is $1 - v_i$, assuming that the probability that E_i occurs is v_i.

It is known, and we just proved this for the given problem, that the independence stated implies the independence of the events

$$E_{i_1}, E_{i_2}, \ldots, E_{i_s}; \bar{E}_{j_1}, \bar{E}_{j_2}, \ldots, \bar{E}_{j_t},$$

assuming that $i_1, i_2, \ldots, i_s; j_1, j_2, \ldots, j_t$ are distinct natural numbers not greater than r. In particular, we have determined the value $\varphi(n)/n$ as the probability that $\bar{E}_1, \bar{E}_2, \ldots, \bar{E}_r$ occur simultaneously.

Exercises:

1. Determine those numbers n for which $\varphi(n)$ divides n.

2. Prove that if a and b are square-free and distinct, then

$$\frac{\varphi(a)}{a} \neq \frac{\varphi(b)}{b}.$$

3. Which numbers a and b satisfy $\varphi(a)/a = \varphi(b)/b$?

4*. Prove that there are infinitely many values n such that the smallest solution to the equation $\varphi(x) = n$ is even.

5. Prove that if $n > 1$, then the sum of the numbers less than n and relatively prime to n is $n\varphi(n)/2$.

4. We mentioned the perfect and the amicable numbers at the very beginning of this book. These are related to the sum of the divisors of a number. Let $\sigma(n)$ denote the sum of the divisors of the number n (counting n itself among its divisors). This is another arithmetic function. Then the perfect numbers are those numbers t that satisfy $\sigma(t) = 2t$, and b and a form an amicable pair if $\sigma(b) = \sigma(a) = b + a$.

Clearly, $\sigma(1) = 1$ and $\sigma(n) \geq n + 1$ for $n > 1$. Equality holds for the prime numbers. We mentioned that already EUCLID stated that $2^n \left(2^{n+1} - 1\right)$ is perfect whenever $2^{n+1} - 1 = p$ is a prime. Indeed, in this case the divisors are

$$1, 2, \ldots, 2^n, p, 2p, \ldots, 2^n p,$$

and their sum is

$$(2^{n+1} - 1)(p + 1) = (2^{n+1} - 1)2^{n+1} = 2 \cdot 2^n (2^{n+1} - 1).$$

5. Let us examine now when an even number can be perfect. Let us write the number as $n = 2^k m$, where $k \geq 1$ and m is an odd number. Let us denote the divisors of m by $1 = m_1, m_2, \ldots, m_r$; their sum is $\sigma(m)$. Then we have

$$\sigma(n) = 1 + 2 + \cdots + 2^k + m_2 + 2m_2 + \cdots + 2^k m_2 + \cdots + m_r$$
$$+ 2m_r + \cdots + 2^k m_r$$
$$= \left(1 + 2 + \cdots + 2^k\right) \sigma(m) = \left(2^{k+1} - 1\right) \sigma(m).$$

We thus have to solve the following equation:

$$\left(2^{k+1} - 1\right) \sigma(m) = 2 \cdot 2^k m = 2^{k+1} m.$$

It is convenient to rewrite this equation as

$$\left(2^{k+1} - 1\right) \left(\sigma(m) - m\right) = m. \tag{1}$$

This means that $\sigma(m) - m = m'$ is a divisor of m, and m' is less than m since $k \geq 1$. We have thus found that $\sigma(m) = m + m'$, i.e., $\sigma(m)$ is the sum of two distinct divisors. The numbers with two distinct divisors are the prime numbers. Their divisors are the number itself and 1; hence $m' = 1$, and from (1) we see that

$$2^{k+1} - 1 = m$$

is a prime number. Furthermore, we prove that this number can be prime only if $k + 1$ is a prime. Indeed, if $k + 1 = uv, u \geq 2, v \geq 2$, then

$$2^{k+1} - 1 = 2^{uv} - 1 = \left(2^u - 1\right) \left(2^{u(v-1)} + 2^{u(v-2)} + \cdots + 2^u + 1\right),$$

and since both factors are greater than 1, the number is composite.

We summarize our results:

Theorem 1 *An even number is perfect if and only if it has the form*

$$2^{p-1}(2^p - 1),$$

and the second term here is a prime number, for which it is necessary that p also be a prime number.

6. If the value of p is $2, 3, 5$, or 7, then the value of $2^p - 1$ is $3, 7, 31, 127$, respectively. These are also prime numbers; hence the four smallest even perfect numbers are $2 \cdot 3 = 6, 4 \cdot 7 = 28, 16 \cdot 31 = 496$, and $64 \cdot 127 = 8128$. The next prime, 11, however, does not lead to a perfect number, since $2^{11} - 1 = 2047 = 23 \cdot 89$ is composite.

The prime numbers among the numbers of the form $M_p = 2^p - 1$ (p prime) are called *Mersenne primes*. A useful criterion aids their study. Let us form the sequence

$$u_1 = 4, \qquad u_{n+1} = u_n^2 - 2 \qquad (n = 1, 2, \dots).$$

For $p \geq 3$, the number M_p is prime if and only if M_p divides u_{p-1}. This criterion, due to LUCAS, is amenable to machine computation, since we only need to compute the residues modulo M_p of the u_n. The number $2^{13\,466\,917} - 1$ was found to be a Mersenne prime on December 7, 2001. This number, (then) the largest known Mersenne prime, has over four million digits and is the 39th known Mersenne prime.[1]

A conjecture belonging to this subject states that the members of the sequence

$$3, 2^3 - 1 = 7, 2^7 - 1 = 127, 2^{127} - 1 = p, 2^p - 1, \dots$$

are all primes. Among these, $2^{127} - 1$ has been known to be prime for nearly a century. For a long time, until the advent of computers, this was the largest known prime (it has a "mere" 38 digits). There was no risk in formulating the conjecture; even with computers, it is hopeless to refute it. (The exponent of 2 in the largest known Mersenne prime given above has only 7 digits.) However, the related conjecture that if $2^p - 1 = q$ is prime for some prime p, then $2^q - 1$ is also a prime was refuted by a computer in the "stone age of computers": The number

$$2^{2^{13}-1} - 1 = 2^{8191} - 1$$

was found to be composite.

Returning to the Mersenne primes, we do not know whether there are infinitely many of them. At the other extreme, the possibility has not been ruled out that among the Mersenne-type numbers of prime index, only finitely many are composite.

7. The proof of Theorem 1 was based on the fact that we could calculate $\sigma(n)$ by determining the product of the divisors of 2^k and m. As one expects, this is not just by chance, since σ is a multiplicative function (in the sense of Section 2.39); i.e., if n_1 and n_2 are relatively prime, then $\sigma(n_1 n_2) = \sigma(n_1)\sigma(n_2)$. To see this, it is enough to show that every divisor d of $n_1 n_2$ can be uniquely written as the product of divisors d_1 of n_1 and d_2 of n_2. The canonical decompositions of n_1 and n_2 are

$$n_1 = p_1^{a_1} p_2^{a_2} \cdots p_r^{a_r}, \qquad n_2 = q_1^{b_1} q_2^{b_2} \cdots q_s^{b_s}.$$

The fact that the two numbers are relatively prime means that every p_i is different from all the q_j's. Hence the canonical decomposition of $n_1 n_2$ is

---

[1] Regarding the Mersenne primes, see, e.g., P. RIBENBOIM: *The Little Book of Big Primes*, Springer-Verlag, Berlin, 1991, pp. 65–72. For information on prime numbers on the World Wide Web, including the current largest prime, see *The Prime Page*: `http://www.utm.edu/research/primes/`

$$n_1 n_2 = p_1^{a_1} p_2^{a_2} \cdots p_r^{a_r} q_1^{b_1} q_2^{b_2} \cdots q_s^{b_s}.$$

All divisors of this number can be written as

$$d = p_1^{\alpha_1} p_2^{\alpha_2} \cdots p_r^{\alpha_r} q_1^{\beta_1} q_2^{\beta_2} \cdots q_s^{\beta_s},$$

where

$$0 \leq \alpha_i \leq a_i, \quad i = 1, 2, \ldots, r; \qquad 0 \leq \beta_j \leq b_j, \quad j = 1, 2, \ldots, s.$$

This has a unique decomposition into divisors of n_1 and n_2:

$$d_1 = p_1^{\alpha_1} p_2^{\alpha_2} \cdots p_r^{\alpha_r}, \qquad d_2 = q_1^{\beta_1} q_2^{\beta_2} \cdots q_s^{\beta_s},$$

because the p's and q's are all distinct. With this we have proved our claim.

On the other hand, if d_1 and d_2 are divisors of n_1 and n_2, respectively, then they uniquely define $d = d_1 d_2$, a divisor of $n_1 n_2$. Thus if we multiply every divisor of n_1 by every divisor of n_2, then on the one hand we have the product of the sum of the divisors of n_1 with the sum of the divisors of n_2, and on the other hand we have the sum of the divisors of the product $n_1 n_2$. We have shown that

$$\sigma(n_1)\sigma(n_2) = \sigma(n_1 n_2),$$

and hence proved the multiplicativity of $\sigma(n)$:

Theorem 2 *The function σ is multiplicative.*

8. It is clear that the value of a multiplicative function f for a product of pairwise relatively prime numbers is equal to the product of the values of f for the factors themselves. Using this argument for the canonical decomposition of a number n, we see that $f(n)$ is equal to the product of f evaluated at the prime powers. Thus in order to determine the value of a multiplicative function, it is enough to calculate it for prime powers.

In the case of σ, if p is a prime and k is a positive integer, then

$$\sigma\left(p^k\right) = 1 + p + p^2 + \cdots + p^k = \frac{p^{k+1} - 1}{p - 1}.$$

Thus if the canonical decomposition of a number n is

$$n = p_1^{k_1} p_2^{k_2} \cdots p_r^{k_r}, \tag{2}$$

then

$$\sigma(n) = \frac{\left(p_1^{k_1+1} - 1\right) \left(p_2^{k_2+1} - 1\right) \cdots \left(p_r^{k_r+1} - 1\right)}{(p_1 - 1)(p_2 - 1) \cdots (p_r - 1)}.$$

Exercises:

6. Prove the following identity:

$$(\sigma(n))^2 = \sum_{d|n} \frac{n}{d} \sigma\left(d^2\right).$$

7. Denote by $\sigma_k(n)$ the sum of the kth powers of the divisors of n, where k is an integer ($\sigma_1(n) = \sigma(n)$, $\sigma_0(n) = \tau(n)$). Give a formula to calculate the value of $\sigma_k(n)$, given the canonical decomposition of n.

8. What relationship exists between $\sigma_k(n)$ and $\sigma_{-k}(n)$?

9. Prove that if there is an odd perfect number, then it is divisible by the square of at least one prime. In fact, in its canonical decomposition, exactly one prime occurs to an odd power, and this prime is of the form $4k + 1$.

10. Prove that $n - \varphi(n)$ is arbitrarily large, and similarly that $\sigma(n) - n$ is arbitrarily large, for n large enough and not prime.

11*. Prove that for any δ greater than 1 and any positive h, there is an n such that $|\sigma(n)/n - \delta| < h$.

12*. Prove that for any δ between 0 and 1, and any positive h, there is an n such that $|\varphi(n)/n - \delta| < h$.

13. Prove that $\varphi(n)\sigma(n) < n^2$, but that there is a positive constant c such that $\varphi(n)\sigma(n) \geq cn^2$ holds for every n.

9. In the following we will consider two new arithmetic functions. We will represent the number of (distinct) prime divisors of n by $\omega(n)$, and the number of prime-power divisors by $\Omega(n)$.[2] For numbers of the form (2),

$$\omega(n) = r, \qquad \Omega(n) = k_1 + k_2 + \cdots + k_r, \qquad \omega(1) = \Omega(1) = 0.$$

It is clear for these two functions that if n_1 and n_2 are relatively prime, then $\omega(n_1 n_2) = \omega(n_1) + \omega(n_2)$, and for any integers n_1 and n_2,

$$\Omega(n_1 n_2) = \Omega(n_1) + \Omega(n_2).$$

Arithmetic functions f that satisfy

$$f(n_1 n_2) = f(n_1) + f(n_2)$$

whenever $(n_1, n_2) = 1$ we will call *additive*, and if equality holds for all n_1 and n_2, we will call f *totally* or *completely additive*. (Similarly, if an arithmetic function g satisfies $g(n_1 n_2) = g(n_1)g(n_2)$ for all n_1 and n_2, we will call g *completely multiplicative*.) Thus the function ω is additive, and Ω is

---

[2] These are sometimes denoted by $\kappa(n)$ and $\nu(n)$, respectively.

completely additive. It is clear that a completely additive function is additive and a completely multiplicative function is multiplicative.

If a multiplicative function f takes on a value other than 0, and $f(a)$ is such a value, then

$$f(a) = f(1 \cdot a) = f(1)f(a); \qquad \text{hence} \qquad (1 - f(1))f(a) = 0,$$

which is possible only if $f(1) = 1$.

If g is an additive function, then for any a,

$$g(a) = g(1 \cdot a) = g(1) + g(a),$$

from which it follows that $g(1) = 0$.

In general, additive and multiplicative functions are not difficult to calculate from the canonical decomposition of a number. The proof of the second part of the following theorem is left to the reader.

Theorem 3 *The value of an additive function at 1 is 0, that of a multiplicative function is 1 if the function is not identically 0.*

If f is an additive function, F is completely additive, g is multiplicative, and G is completely multiplicative, then for n of the form (2), we have

$$f(n) = f\left(p_1^{k_1}\right) + f\left(p_2^{k_2}\right) + \cdots + f\left(p_r^{k_r}\right),$$
$$F(n) = k_1 F(p_1) + k_2 F(p_2) + \cdots + k_r F(p_r),$$
$$g(n) = g\left(p_1^{k_1}\right) g\left(p_2^{k_2}\right) \cdots g\left(p_r^{k_r}\right),$$
$$G(n) = G(p_1)^{k_1} \cdot G(p_2)^{k_2} \cdots \cdots G(p_r)^{k_r}.$$

It follows from the theorem that no matter how we assign values to a function for all prime powers, this uniquely determines an additive function as well as a multiplicative function. Completely additive and completely multiplicative functions are determined by their values for the primes.

Naturally, the classes of additive and multiplicative functions do not include all arithmetic functions. For example, $\pi(n)$ belongs to neither class.

Exercises:

14. Prove that if $k > 1$ is an integer, then neither $g(n) = \tau(kn)$ nor $h(n) = \sigma(kn)$ is multiplicative.

15. Prove that if f and g are multiplicative arithmetic functions, $f \neq g$, and neither is identically 0, then neither $h(n) = f(n) + g(n)$ nor $d(n) = f(n) - g(n)$ is multiplicative.

16. If g and h are multiplicative functions, is it possible that the function $h(n) = (f(n) + g(n))/2$ is multiplicative?

17. Prove that if f and g are additive arithmetic functions, and for n greater than 1, $f(n)$ and $g(n)$ are positive, then $h(n) = f(n)g(n)$ is not additive. Prove that if we omit the condition of their being positive, then the conclusion does not necessarily hold.

10. We call those numbers perfect for which $\sigma(n)/n = 2$. It is customary to call numbers n for which $\sigma(n)/n > 2$ *abundant* and those for which $\sigma(n)/n < 2$ *deficient*.

We see, for example, that every prime power is deficient. If p is a prime, then

$$\frac{\sigma\left(p^k\right)}{p^k} = \frac{p^{k+1}-1}{p^k(p-1)} < \frac{p^{k+1}}{p^k(p-1)} = \frac{p}{p-1} = 1 + \frac{1}{p-1} \le 2.$$

It is also true that the product of two odd prime powers is deficient. If the two primes are $2 < p < q$, then

$$\frac{\sigma(p^j q^k)}{p^j q^k} = \frac{p^{j+1}-1}{p^j(p-1)} \cdot \frac{q^{k+1}-1}{q^k(q-1)} < \frac{p}{p-1} \cdot \frac{q}{q-1}$$

$$= \left(1 + \frac{1}{p-1}\right)\left(1 + \frac{1}{q-1}\right) \le \left(1 + \frac{1}{2}\right)\left(1 + \frac{1}{4}\right) = \frac{15}{8} < 2.$$

It is also easy to see that for a number n, the value $\sigma(n)/n$ is always smaller than that of its multiples. If the divisors of an arbitrary natural number n are $n_1, n_2, \ldots, n_\tau$, and a is an integer greater than 1, then $an_1, an_2, \ldots, an_\tau$ are divisors of an, all greater than 1. Thus

$$\sigma(an) \ge an_1 + an_2 + \cdots + an_\tau + 1 = a\sigma(n) + 1,$$

and hence

$$\frac{\sigma(an)}{an} \ge \frac{\sigma(n)}{n} + \frac{1}{an},$$

proving the claim.

11*. We have already encountered even perfect numbers, and in fact we know that every even perfect number has exactly two prime divisors. We do not know any odd perfect numbers, but if any exist, we can show something about the number of prime divisors.

Theorem 4 *For an arbitrary s, there are at most finitely many odd perfect numbers with exactly s distinct prime divisors.*

We will prove the theorem indirectly. During the proof we will make use of the laws of calculations of limits.

Let us assume that for some s there are infinitely many odd perfect numbers with s prime divisors. Then there would be primes that divide only finitely many of the perfect numbers under consideration, and there would be primes that occur in the canonical decomposition of infinitely many. For the latter primes, there are those that occur infinitely often to the same power in the canonical decomposition of our perfect numbers, and those that do not.

Let p_1 be a prime—if such exists—that occurs infinitely often with the same power, say k_1. Then let us consider only those perfect numbers for which p_1 occurs to the power k_1 in the canonical decomposition. If among these there are infinitely many for which p_2 occurs to the power k_2 in the canonical decomposition, then we consider only those. We repeat this procedure as many times as possible.

This must stop after finitely many steps, since these numbers have only s prime divisors, and we are left with infinitely many odd perfect numbers such that $p_1^{k_1} p_2^{k_2} \cdots p_u^{k_u}$ occurs in the canonical decomposition and no other prime occurs infinitely often to the same power. It is still possible that infinitely many of the remaining odd perfect numbers are divisible by a prime q_1. If this is the case, we consider only those divisible by q_1. If infinitely many of the remaining numbers are divisible by q_2, then again we only retain those.

This also stops after finitely many steps. The infinitely many odd perfect numbers $n_1, n_2, \ldots$ that remain have canonical decomposition

$$n_i = p_1^{k_1} p_2^{k_2} \cdots p_u^{k_u} q_1^{\ell_{i1}} q_2^{\ell_{i2}} \cdots q_v^{\ell_{iv}} r_{i1}^{m_{i1}} r_{i2}^{m_{i2}} \cdots r_{iw}^{m_{iw}} \tag{3}$$

$$(u + v + w = s; \quad i = 1, 2, \ldots).$$

All of the perfect numbers under consideration are different, so it is not possible that $v = w = 0$; among the exponents, $\ell_{i1}, \ell_{i2}, \ldots, \ell_{iv}$ as well as among the primes, $r_{i1}, r_{i2}, \ldots, r_{iw}$, no value can appear infinitely often. The values of ℓ_{ij} and r_{ij} can be smaller than a given bound C only finitely many times, and hence every ℓ_{ij} and every r_{ij} is larger than C for i large enough.

Let us now write what it means for n_i to be perfect:

$$2 = \frac{\sigma(n_i)}{n_i} = \frac{p_1^{k_1+1} - 1}{p_1^{k_1}(p_1 - 1)} \cdot \frac{p_2^{k_2+1} - 1}{p_2^{k_2}(p_2 - 1)} \cdots \frac{p_u^{k_u+1} - 1}{p_u^{k_u}(p_u - 1)}$$

$$\frac{q_1^{\ell_{i1}+1} - 1}{q_1^{\ell_{i1}}(q_1 - 1)} \cdot \frac{q_2^{\ell_{i2}+1} - 1}{q_2^{\ell_{i2}}(q_2 - 1)} \cdots \frac{q_v^{\ell_{iv}+1} - 1}{q_v^{\ell_{iv}}(q_v - 1)}$$

$$\frac{r_{i1}^{m_{i1}+1} - 1}{r_{i1}^{m_{i1}}(r_{i1} - 1)} \cdot \frac{r_{i2}^{m_{i2}+1} - 1}{r_{i2}^{m_{i2}}(r_{i2} - 1)} \cdots \frac{r_{iw}^{m_{iw}+1} - 1}{r_{iw}^{m_{iw}}(r_{iw} - 1)}.$$

12*. The first u factors here are independent of i. In the next v factors, the term q_j in the denominator is independent of i, and we can rewrite these factors as

$$\frac{q_j^{\ell_{ij}+1} - 1}{q_j^{\ell_{ij}}(q_j - 1)} = \frac{q_j - 1/q_j^{\ell_{ij}}}{q_j - 1}, \qquad j = 1, 2, \ldots, v.$$

If i grows without bound, then so does ℓ_{ij}, and thus the second term in the numerator tends to 0, and the value of the quotient tends to

$$\frac{q_j}{q_j - 1}.$$

The third type of factors can be written as

$$\frac{r_{ij}^{m_{ij}+1} - 1}{r_{ij}^{m_{ij}}(r_{ij} - 1)} = \frac{1 - 1/r_{ij}^{m_{ij}+1}}{1 - 1/r_{ij}}, \qquad j = 1, 2, \ldots, w.$$

Here, r_{ij} grows without bound as i does, and thus the numerator, the denominator, and hence the quotient, all tend to 1. Because the product of terms from finitely many convergent sequences is also convergent, and it converges to the product of the limits of the factors, it follows that if i grows beyond all bounds, then

$$\frac{\sigma(n_i)}{n_i} \to \frac{p_1^{k_1+1} - 1}{p_1^{k_1}(p_1 - 1)} \cdot \frac{p_2^{k_2+1} - 1}{p_2^{k_2}(p_2 - 1)} \cdots \frac{p_u^{k_u+1} - 1}{p_u^{k_u}(p_u - 1)} \cdot \frac{q_1}{q_1 - 1} \cdot \frac{q_2}{q_2 - 1} \cdots \frac{q_v}{q_v - 1}.$$

The value of the fraction on the left is 2 for every i, and hence its limit is also 2. It follows that v cannot be 0, since this would mean that $p_1^{k_1} p_2^{k_2} \cdots p_u^{k_u}$ is a perfect number; each of its multiples would be abundant, and hence the sequence of n_i would have only one term.

Multiplying both sides by $p_1^{k_1} p_2^{k_2} \cdots p_u^{k_u}(q_1 - 1)(q_2 - 1) \cdots (q_v - 1)$, we obtain the following identity:

$$\frac{p_1^{k_1+1} - 1}{(p_1 - 1)} \cdot \frac{p_2^{k_2+1} - 1}{(p_2 - 1)} \cdots \frac{p_u^{k_u+1} - 1}{(p_u - 1)} \cdot q_1 q_2 \cdots q_v$$
$$= 2 p_1^{k_1} p_2^{k_2} \cdots p_u^{k_u}(q_1 - 1)(q_2 - 1) \cdots (q_v - 1).$$

Every factor on the left-hand side is an integer, and thus the right-hand side is divisible by the q's. If q_1 is the largest of the q's, it is odd, distinct from the p_i's, and does not divide the $q_j - 1$, because it is prime and larger. This is a contradiction, and Theorem 4 must therefore be true.

The theorem is due to DICKSON.[3] The simple proof presented above was found by RÓZSA PÉTER as a university student, unaware of Dickson's result; she did not publish it.

13*. Close relatives of perfect numbers are the *primitive abundant* numbers. These are numbers that are not deficient, but all of whose divisors are. Perfect numbers are examples, but there are infinitely many other primitive abundant numbers as well. Starting from an arbitrary abundant number, we discard

---

[3] L. E. DICKSON: *Amer. Journ. of Math.* **35** (1913), pp. 413–422. For a proof similar to that presented above, see: H. N. SHAPIRO: *Bull. Amer. Math. Soc.*, **55** (1949), 450–452.

a prime divisor. If the quotient is still abundant, we discard another prime divisor. After finitely many steps, we arrive at a deficient number, at the latest when only a prime power remains. This procedure can stop only when we arrive at a primitive abundant number. Examples of primitive abundant numbers are $20, 70, 88, 3^3 \cdot 5 \cdot 7 = 945$.

The proof from the previous section can be applied, almost without changes, to primitive abundant numbers having s prime divisors. If there were infinitely many of these, then there would be an infinite sequence of the form (3), with $v + w \geq 1$. The limit of the sequence $\sigma(n_i)/n_i$ is still

$$\frac{p_1^{k_1+1} - 1}{p_1^{k_1}(p_1 - 1)} \cdot \frac{p_2^{k_2+1} - 1}{p_2^{k_2}(p_2 - 1)} \cdots \frac{p_u^{k_u+1} - 1}{p_u^{k_u}(p_u - 1)} \cdot \frac{q_1}{q_1 - 1} \cdot \frac{q_2}{q_2 - 1} \cdots \frac{q_v}{q_v - 1}.$$

On the other hand, because n_i is a primitive abundant number it follows that $\sigma(n_i)/n_i \geq 2$; if $v \geq 1$, then

$$\frac{\sigma(n_i/q_1)}{n_i/q_1} < 2,$$

and if $w \geq 1$, then

$$\frac{\sigma(n_i/r_{i1})}{n_i/r_{i1}} < 2.$$

In the first case,

$$2 \leq \frac{\sigma(n_i)}{n_i} = \frac{\sigma(n_i/q_1)}{n_i/q_1} \cdot \frac{\sigma(n_i)}{q_1 \sigma(n_i/q_1)} < 2 \cdot \frac{\frac{q_1^{\ell_{i1}+1} - 1}{q_1 - 1}}{\frac{q_1\left(q_1^{\ell_{i1}} - 1\right)}{q_1 - 1}} = 2\left(1 + \frac{1 - 1/q_1}{q_1^{\ell_{i1}} - 1}\right).$$

In the second case,

$$2 \leq \frac{\sigma(n_i)}{n_i} = \frac{\sigma(n_i/r_{i1})}{n_i/r_{i1}} \cdot \frac{\sigma(n_i)}{r_{i1} \sigma(n_i/r_{i1})} < 2 \cdot \frac{\frac{r_{i1}^{m_{i1}+1} - 1}{r_{i1} - 1}}{\frac{r_{i1}\left(r_{i1}^{m_{i1}} - 1\right)}{r_{i1} - 1}} = 2\left(1 + \frac{1 - 1/r_{i1}}{r_{i1}^{m_{i1}} - 1}\right).$$

The right-hand summand in parentheses tends to 0 in both cases, hence the limits are both 2. The proof proceeds in a fashion similar to that of perfect numbers.

14. Let us now consider the sum of the reciprocals of primitive abundant numbers up to some integer. ERDŐS showed that this does not grow beyond a certain bound.[4] From this fact and Exercise 6.15*, it follows that the integers not divisible by primitive abundant numbers have asymptotic density, and thus so do the numbers not included, i.e., those divisible by a primitive abundant number. The latter are the abundant numbers; the former are

---

[4] P. ERDŐS: *Journal London Math. Soc.* **9** (1934), pp. 278–282.

the deficient numbers. DAVENPORT, CHOWLA, and BEHREND proved in a different manner that abundant numbers have asymptotic density.[5]

From the point of view of density, the perfect numbers do not play much of a role. Even if we consider the so-called k-fold perfect numbers, those for which $\sigma(n)/n$ is an integer k, we do not even know whether there are infinitely many of these. MERSENNE, DESCARTES, and FERMAT already knew of many examples of multiple-fold perfect numbers, for example $\sigma(672) = 3 \cdot 672$ (FERMAT), $\sigma(1\,476\,304\,896) = 3 \cdot 1\,476\,304\,896$ (DESCARTES). DESCARTES determined six 4-fold perfect numbers and one 5-fold perfect number. To date, 334 multiple-fold perfect numbers are known. The values of k for those known go up to 8. Let $n_1 = 6 < n_2 < n_3 < \cdots$ be the sequence of all perfect and all multiple-fold perfect numbers, and let $A(x)$ be the numbers of terms not greater than x. We do not even know whether this sequence has finitely many terms, i.e., whether A is bounded.

The result mentioned above states that this sequence has density 0. HORNFECK and WIRSING, sharpening results of KANOLD, ERDŐS, and others, were able to prove in a surprisingly easy way the fact that for any positive h, the quotient $A(x)/x^h$ is arbitrarily small for x large enough.[6] It is a special case of this (when $h = 1$) that the sequence has density 0. WIRSING later improved this result, showing that for some positive constant c,

$$A(x) < x^{c\sqrt{\log x}}.$$

Further sharpening of this appears quite difficult, although the correct order of magnitude of $A(x)$ is probably much smaller.

Denote by $N_2(x)$ the number of primitive abundant numbers not greater than x. The following bounds are known:[7]

$$\frac{x}{\exp\left(\sqrt{2}\sqrt{\log x \log \log x}\right)} < N_2(x) < \frac{x}{\exp\left(\sqrt{\log x \log \log x}\right)}.$$

Let us call a number n α-abundant if $\sigma(n)/n > \alpha$, and primitive α-abundant if $\sigma(n)/n \geq \alpha$ and $\sigma(d)/d < \alpha$ for every proper divisor d of n. The quotient of $N_\alpha(x)$, the number of α-abundant numbers not greater than x, and $x/\log x$ is arbitrarily small, for x large enough. On the other hand, if $g(x)$ is any slowly growing unbounded function, then there are an α and infinitely many x such that

$$\frac{N_\alpha(x)g(x)}{x/\log x}$$

remains larger than some positive bound.[8]

---

[5] H. DAVENPORT: *Sitzungsberichte der Preussischen Akademie der Wissenschaften* **27** (1933), pp. 830–837, F. BEHREND: *ibid.* **22** (1932), pp. 322–328, and **6** (1933), pp. 280–293.

[6] B. HORNFECK and E. WIRSING: *Math. Annalen* **133** (1957), pp. 431–438.

[7] M. AVIDON: *Acta Arithmetica* **77** (1996), pp. 195–205.

[8] P. ERDŐS: *Acta Arithmetica* **5** (1959), pp. 25–33.

ERDŐS further showed that there is an α for which the sum of the reciprocals of primitive α-abundant numbers grows beyond all bounds. BEHREND, DAVENPORT, and CHOWLA showed that the sequence of α-abundant numbers has asymptotic density for every α greater than 1. We denote this density by $f(\alpha)$, and this function is continuous and monotone decreasing in α. It is obvious that $f(1) = 1$, and it can be shown that $f(\alpha)$ can be made arbitrarily small for α large enough. The first result of a similar type was by SCHOENBERG who showed for the function $\varphi(n)$, whose behavior is somewhat similar to that of $\sigma(n)$, that for every positive c, the sequence of numbers satisfying $\varphi(n)/n \le c$ has asymptotic density.[9] Many results of this type can be found in the literature.[10]

From the results of BEHREND, DAVENPORT, and CHOWLA, it follows easily (however, we will not go into the details here) that the sequence of perfect numbers has density 0.

At the beginning of this chapter we mentioned the amicable numbers. As is the case with perfect numbers, we do not know whether there are infinitely many, although it is likely that there are. ERDŐS was able to show that the sequence of numbers each of which is in at least one amicable pair has density 0.[11]

Exercises:

18. Prove that deficient numbers have deficient multiples.

19. Prove that there are infinitely many odd primitive abundant numbers.

20. Prove that there are infinitely many primitive abundant numbers that have only one odd prime divisor.

21. Prove that odd numbers having at most 6 distinct prime factors, and not divisible by 3 are deficient.

Denote by $e(n)$ the function that is always 1, and by $u(n)$ the function given by $u(n) = n$. The functions $\tau(n)$ and $\sigma(n)$ each occur as the sum of these functions, respectively, taken over all divisors of n. In general, if f is an arithmetic function, then the function F defined by $F(n) = \sum_{d|n} f(d)$ is called the *summatory function* of f. We saw that Euler's φ-function is the summatory function of $u(n)$, and that this uniquely determines φ (Theo-

---

[9] I. J. SCHOENBERG: *Math. Zeitschr.* **98** (1928), pp. 171–199.

[10] See, for example, M. KAC: *Bull. Amer. Math. Soc.* **55** (1949), pp. 641–665; P. ERDŐS: *Proc. Internat. Cong. Mat.*, Amsterdam, Groningen, E. P. Noordhoff, 1954, Vol. III, pp. 13–19; I. P. KUBILIUS: *Uspechi Math. Nauk.* (N.S.) **11** (1956), no. 2 (68) pp. 31–66 (in Russian), and *Amer. Math. Transl.*, (2) **19** (1962), pp. 47–85.

[11] P. ERDŐS: *Publicationes Math. Debrecen* **4** (1955), pp. 108–111. See also: I. Z. RUZSA: *Mat. Lapok,* **27** (1976–1979), pp. 95–143 and pp. 227–283 (in Hungarian), and P. D. T. A. ELLIOT: *Probabilistic Number Theory* I–II., Springer-Verlag, Berlin, 1979.

rem 2.9). Is it possible to determine the original function from its summatory function? The following problems will deal with this question.

22. Is it true that the summatory function of a multiplicative function is multiplicative?

23. Determine those completely multiplicative functions whose summatory function is also completely multiplicative.

24. Determine the summatory function of the so-called Möbius function $\mu(n)$ defined by $\mu(1) = 1$, $\mu(n) = (-1)^r$ if n is the product of r distinct primes, and $\mu(n) = 0$ otherwise (when n is divisible by a square greater than 1).

25. Prove that if F is the summatory function of f, then

$$f(n) = \sum_{d|n} \mu(d)F\left(\frac{n}{d}\right). \tag{4}$$

26. What is the summatory function of the function that is defined in terms of F in (4)?

For arbitrary functions f and g, we may define a new function h in a fashion similar to (4):

$$h(n) = \sum_{d|n} f(d)g\left(\frac{n}{d}\right).$$

The function h is called the *Dirichlet product* or *convolution* of f and g, and is denoted by $h(n) = (f * g)(n)$. The summatory function is then the convolution with $e(n)$.

27. Is convolution
   (a) commutative?
   (b) associative?
   (c) distributive for addition of two functions?

28. Find an identity function for convolution, i.e., a function $i(n)$ for which $i * f = f$ for all functions f.

29. Which functions f have an inverse for convolution, i.e., a function f' such that $f * f' = i$?

15. We saw that multiplicative and additive arithmetic functions are determined by arbitrarily assigning values for all prime powers; even completely multiplicative and completely additive functions can be arbitrarily assigned values for primes. In this way, we see that we can create functions that behave in a variety of ways. Yet already the functions we have encountered so far,

$$\tau(n), \quad \omega(n), \quad \Omega(n), \quad \sigma(n), \quad \varphi(n),$$

are quite diverse. For a prime p, we have

$$\tau(p) = 2, \qquad \omega(p) = \Omega(p) = 1, \qquad \sigma(p) = p + 1, \qquad \varphi(p) = p - 1;$$

if k is a positive integer, then

$$\tau\left(p^k\right) = k + 1, \qquad \omega\left(p^k\right) = 1, \qquad \Omega\left(p^k\right) = k,$$

$$\sigma\left(p^k\right) = \frac{p^{k+1} - 1}{p - 1}, \qquad \varphi\left(p^k\right) = p^k - p^{k-1} = p^{k-1}(p - 1).$$

Let $N = 2 \cdot 3 \cdot p_3 \cdots p_s$ be the product of the first s primes, and consider these functions for N:

$$\tau(N) = 2^s, \quad \omega(N) = \Omega(N) = s, \quad \sigma(N) = (2 + 1)(3 + 1)(p_3 + 1) \cdots (p_s + 1),$$

$$\varphi(N) = (2 - 1)(3 - 1)(p_3 - 1) \cdots (p_s - 1).$$

This already shows that the functions $\tau(n), \omega(n), \Omega(n)$ are small (either 1 or 2) infinitely often, yet also take on arbitrarily large values; frequently $\tau(n)$ and $\omega(n)$ are close in value, as are $\sigma(n)$ and $\varphi(n)$, yet $\tau(n)$ can also be "very large" compared to $\omega(n)$.

We see that $\sigma(n) \geq n + 1, \varphi(n) \leq n - 1$ (with equality holding for the primes), yet for N defined above, $\sigma(N)$ is "significantly larger" than N, and $\varphi(N)$ is "significantly smaller" than N. More precisely, $\sigma(N)/N$ can be arbitrarily large, and $\varphi(N)/N$ arbitrarily small, if s is large enough. If $s > 2$, then using Theorem 5.4, it follows that

$$\frac{\sigma(N)}{N} = \left(1 + \frac{1}{2}\right)\left(1 + \frac{1}{3}\right)\left(1 + \frac{1}{p_3}\right) \cdots \left(1 + \frac{1}{p_s}\right)$$

$$> 1 + \frac{1}{2} + \frac{1}{p_3} + \cdots + \frac{1}{p_s} > \log\log p_s.$$

On the other hand, with the help of the inequality for the exponential function in Fact 5, we have

$$\frac{\varphi(N)}{N} = \left(1 - \frac{1}{2}\right)\left(1 - \frac{1}{3}\right)\left(1 - \frac{1}{p_3}\right) \cdots \left(1 - \frac{1}{p_s}\right)$$

$$< \exp\left(-\frac{1}{2}\right) \exp\left(-\frac{1}{3}\right) \exp\left(-\frac{1}{p_3}\right) \cdots \exp\left(-\frac{1}{p_s}\right)$$

$$= \exp\left(-\frac{1}{2} - \frac{1}{3} - \frac{1}{p_3} - \cdots - \frac{1}{p_s}\right) < \exp(-\log\log p_s + 1).$$

This proves our claims.

These "small" and "large" values can occur next to each other. According to Dirichlet's theorem (Section 2.19) for N defined above, there are infinitely many primes of the form $Nk + 1$. Let p be such a prime. Then for $p - 1$,

$$\tau(p - 1) \geq \tau(N), \qquad \omega(p - 1) \geq \omega(N), \qquad \Omega(p - 1) \geq \Omega(N).$$

The same inequality is also true for the function $\varphi(n)$, because $\varphi(p-1) = \varphi(Nk)$, and the latter arises from $\varphi(N)$ in such a way that if $k' = (k, N)$ is the largest common divisor of k and N, and $k = k_1 k'$, then $\varphi(kN) = \varphi(k_1)k'\varphi(N)$ (see Exercise 30). Thus

$$\varphi(p-1) \geq \varphi(N).$$

On the other hand, $\varphi(p-1)/(p-1)$ will be small,

$$\frac{\varphi(p-1)}{p-1} = \frac{\varphi(N)}{N} \cdot \frac{\varphi(k_1)}{k_1} \leq \frac{\varphi(N)}{N},$$

whereas $\varphi(p)/p = 1 - 1/p$ is "close" to 1.

It is also not difficult to show that $\sigma(p-1)/(p-1)$ is "large" (whereas $\sigma(p)/p = 1 + 1/p$ is "close" to 1).

Exercise:

30. Prove that if $k' = (k, N)$ is the largest common divisor of k and N, and $k = k_1 k'$, then $\varphi(kN) = \varphi(k_1)k'\varphi(N)$.

16*. The bounds given above are quite large, and in fact, the function $\tau(n)$ actually behaves quite rhapsodically within them. We saw that it takes on very large and very small values for neighboring numbers. We will now show that it can in fact take on values larger than the product of the values for many neighboring values. More precisely, we have the following result:

Theorem 5 *For every natural number k, there is an n such that*

$$\tau(n) > \prod_{j=1}^{k} \tau(n-j)\tau(n+j).$$

We will again consider the product of the first s primes, a number that has often proved useful whenever we needed a number with many divisors. We will determine an appropriate value of s later. Thus let us start with the number

$$N = 2 \cdot 3 \cdot p_3 \cdots p_s.$$

We should not expect that all of the numbers

$$N-k, \quad N-k+1, \quad \ldots, \quad N-1, \quad N, \quad N+1, \quad \ldots, \quad N+k-1, \quad N+k$$

have few divisors, but we will show that if we consider sufficiently many multiples of N arising from a large enough s, then among these, there is one such that none of the k numbers neighboring to the right and left will have many divisors.

Let us examine the neighbors of the numbers $N, 2N, \ldots, (N-1)N$. If s is large enough, then $N > k$, and thus the numbers $rN + t$ ($r = 1, 2, \ldots, N-1$;

$t = \pm 1, \pm 2, \ldots, \pm k$) are all positive and smaller than N^2. For a given value of t, we need to consider the arithmetic progression of $N - 1$ elements. Every one is divisible by $(N, t) = t_1$. Let $N = t_1 N'$, $t = t_1 t'$. Then

$$(N', t') = 1, \qquad rN + t = t_1(rN' + t').$$

The number $rN + t$ can be decomposed at least one way into the product of divisors of t_1 and of $rN' + t'$. (The decomposition will be unique only when $(t_1, rN' + t') = 1$.) Adding together the number of divisors for the terms of the sequence, we have

$$\sum_{r=1}^{N-1} \tau(rN + t) \le \tau(t_1) \sum_{r=1}^{N-1} \tau(rN' + t').$$

The numbers $rN' + t'$ are also an arithmetic progression. Because $(N', t') = 1$, these numbers can only have divisors that are relatively prime to N'. Such a divisor u will divide every uth term of the sequence; the first term divisible by u will occur among the first u terms, and thus the number of terms divisible by u is at most

$$\left[\frac{N-1}{u}\right] + 1 < \frac{N}{u} + 1.$$

Again, we will count the divisors in pairs (whose product is $rN' + t'$); the smaller of the pair is at most

$$\sqrt{rN' + t'} \le \sqrt{rN + t} < N.$$

Rather than taking all numbers relatively prime to N', we consider all numbers less than N, and using the inequality in Fact 6 we obtain

$$\sum_{r=1}^{N-1} \tau(rN' + t') < 2 \sum_{u=1}^{N-1} \left(\frac{N}{u} + 1\right)$$
$$= 2N\left(1 + \frac{1}{2} + \frac{1}{3} + \cdots + \frac{1}{N-1}\right) + 2(N - 1)$$
$$< 2N(1 + \log(N - 1)) + 2N < 2N \log N + 4N,$$

and because $\tau(t_1) < t_1 < k$, it follows that

$$\sum_{r=1}^{N-1} \tau(rN + t) < 2\tau(t_1)(N \log N + 2N) < 3\tau(t_1)N \log N < 3kN \log N. \quad (5)$$

17*. We are looking for an r for which

$$\tau(rN) \geq \prod_{r=1}^{k} \tau(rN - t)\tau(rN + t).$$

We have seen that

$$\tau(rN) \geq \tau(n) = 2^s,$$

and thus the desired inequality will hold if for some fixed s we can show that

$$\tau(rN + t) < 2^{s/2k}, \tag{6}$$

for $t = \pm 1, \pm 2, \ldots, \pm k$. This must be true if among all values $rN + t$ ($r = 1, 2, \ldots, N - 1$; $t = \pm 1, \pm 2, \ldots, \pm k$), there are fewer than $N - 1$ for which (6) does not hold. Thus we have reduced the problem to that of finding an s for which (6) fails in fewer than $N - 1$ cases.

18*. For a fixed t, (5) shows that (6) can fail for at most

$$\frac{3kN \log N}{2^{s/2k}}$$

values. Because t can take on $2k$ possible values, at most

$$\frac{6k^2 N \log N}{2^{s/2k}}$$

numbers can have more that $2^{s/2k}$ divisors. Thus we need to choose s, if possible, so that for $N = 2 \cdot 3 \cdot p_3 \cdots p_s$, the inequalities

$$\frac{6k^2 N \log N}{2^{s/2k}} < N, \qquad \text{i.e.,} \qquad \frac{6k^2 \log N}{2^{s/2k}} < 1,$$

hold. According to Theorem 5.6,

$$N = \prod_{p \leq p_s} p < 4^{p_s}.$$

According to the lower bound in Theorem 5.5, there is a constant c_1 such that

$$s = \pi(p_s) > \frac{c_1 p_s}{\log p_s}.$$

Putting these together we have

$$\frac{6k^2 \log N}{2^{s/2k}} < \frac{(6k^2 \log 4)p_s}{2^{c_1 p_s/2k \log p_s}}.$$

Thus we only need to prove that the logarithm of the right-hand side is negative, or equivalently, that the logarithm of the reciprocal of the right-hand side is positive. For a fixed k, both $c_1 \log 2/(2k)$ and $6k^2 \log 4$ are constants, and we shall denote them by c_2 and c_3, respectively, and we need to find an s such that

$$\frac{c_2 p_s}{\log p_s} - \log p_s - \log c_3 > 0,$$

or equivalently,

$$p_s > \frac{1}{c_2} \log p_s (\log p_s + \log c_3).$$

This is true for every s large enough, because p_s is unbounded as s grows, and by Fact 7, it follows that for p_s large enough, it is larger than any power of $\log p_s$.

With this we have proved Theorem 5.

19. It is not uninteresting to pay attention to the statistical aspects of the proof. The basic idea is that not many of the given numbers can have more divisors than the average number of divisors of the numbers chosen. We thus see that our proof gives no idea how to find an n satisfying the claim of Theorem 5.

The theorem can be sharpened: in place of k we may write a number that grows beyond all bounds with n.[12]

The proof above gave a method that may be refined to give more applications so that instead of N we take many numbers each having a large number of prime divisors, and find multiples of them which are near each other, but such that all numbers close by have few prime divisors. In this way, with minor changes, the method can be applied to many properties of arithmetic functions.

It can be shown, for example, that for any positive t and any (small) positive h, there is an n for which the quotient $\sigma(n+1)/\sigma(n)$ differs by at most h from t. The same holds for $\varphi(n+1)/\varphi(n)$ and $\Omega(n+1)/\Omega(n)$. It is also true that for $(\sigma(n+1) - \sigma(n))/n$ and for $(\varphi(n+1) - \varphi(n))/n$, the statement holds for $-1 < t < 1$.

At the same time, it is unknown whether the analogous statement is true for $\tau(n+1)/\tau(n)$ for every positive t; in fact, with the exception of 0, no values of t are known to which the quotient is arbitrarily close infinitely often, although we know that such values of t exist; in fact, the collection of them is, in some well-defined sense, significant.[13]

It can be shown that the density of the solutions to the inequality $\tau(n+1) \geq \tau(n)$ is $\frac{1}{2}$, and that the density of the solutions of $\tau(n+1) \leq \tau(n)$ is also $\frac{1}{2}$. It is unknown, however, whether there are infinitely many values of n for which $\tau(n+1) = \tau(n)$. Similarly, the density of the solutions of $\sigma(n+1) \geq \sigma(n)$ is $\frac{1}{2}$, as well as for $\sigma(n+1) \leq \sigma(n)$, but it is still unknown whether there are infinitely many solutions to $\sigma(n+1) = \sigma(n)$; and similarly for the equations $\varphi(n+1) = \varphi(n)$ and $\omega(n+1) = \omega(n)$. It is known that $|\omega(n+1) - \omega(n)|$ takes a certain value infinitely often. No similar result is known for the other functions mentioned.

---

[12] ERDŐS P.: *Mat. Lapok* **11** (1960), pp. 26–33 (in Hungarian).
[13] P. ERDŐS: *Acta Arithmetica* **4** (1958), pp. 10–17.

The equation $\varphi(n) = \varphi(n+1) = \varphi(n+2)$ has a solution smaller than $10\,000$, but it is unknown whether for every k there is an n satisfying

$$\varphi(n) = \varphi(n+1) = \varphi(n+2) = \cdots = \varphi(n+k).$$

Analogous statements are also unknown for $\sigma(n)$ and $\tau(n)$.

20. If we consider solutions to the equation $\varphi(m) = \varphi(n)$, but do not require that n and m should be consecutive as we did above, then there are some nice results. For any positive c, it is possible to give sequences m_k and n_k ($k = 1, 2, \ldots$) for which $\varphi(m_k) = \varphi(n_k)$, and m_k/n_k is arbitrarily close to c, for k large enough; in fact, the following even sharper statement is true: For every positive h, every integer $r > 1$, and every $c > 1$, there is a sequence $n_1, n_2, \ldots, n_r$ for which $n_{k+1} < (1+h)n_i$ ($i = 1, 2, \ldots, r-1$), $n_r > cn_1$, and[14]

$$\varphi(n_1) = \varphi(n_2) = \cdots = \varphi(n_r).$$

(The previous statement is implied by this statement by running h through a sequence tending to 0 and selecting n_r and n_1 for each h.)

It is known that there are infinitely many pairs m_k, n_k for which $\varphi(m_k) = \varphi(n_k)$, and $(m_k, n_k) = 1$, but it is not known whether there are infinitely many relatively prime triples m_k, n_k, r_k for which $\varphi(m_k) = \varphi(n_k) = \varphi(r_k)$. It is also unknown whether the sequence of all numbers m for which there is a number n satisfying $\varphi(m) = \varphi(n)$ and $(m, n) = 1$ has density. If we omit the condition of being relatively prime, then it is easy to see that the numbers m satisfying the other condition have density, and in fact this density is 1, showing that the density of numbers occurring as values of $\varphi(n)$ is 0. We will prove this in Theorem 6 below. For relatively prime pairs m and n, the answer is most likely the same; the difficulty arises in that there could be an a for which there are many n_i satisfying $\varphi(n_i) = a$, but no two of which are relatively prime. We need further information about the number of such a's and n_i's.

It is interesting that for the function $\sigma(n)$ we know much less in this direction. There exist sequences of pairs m_k, n_k ($k = 1, 2, \ldots$) for which $\sigma(m_k) = \sigma(n_k)$ and m_k/n_k is arbitrarily close to 1, for k large enough, but no one has been able to prove the corresponding statement if we replace 1 by any arbitrary positive number c. For those values of c for which there exist integers a and b satisfying $c = \sigma(a)/\sigma(b)$, the statement follows rather easily.

No one has succeeded yet in proving that there are sequences m_k, n_k for which $\sigma(m_k) = \sigma(n_k)$ and m_k/n_k grows beyond all bounds. Nor is it known whether there are infinitely many values taken by both $\sigma(n)$ and $\varphi(n)$, although it is very probable that there are. The statement would follow if there are infinitely many twin primes $p, p+2$, because then $\sigma(p) = p+1 = \varphi(p+2)$.

---

[14] P. ERDŐS: Problem 97, *Mat. Lapok* **8** (1957), p. 142, *ibid.* **11** (1960), pp. 152–154 (in Hungarian).

21*. It is clear that the functions $\tau(n), \omega(n), \Omega(n)$ take on all values; in fact, they take on all values—with the exception of the value 1 for τ—infinitely many times. This is no longer true for $\varphi(n)$, which—with the exceptions of $n = 1$ and 2—takes only even values. Indeed, if $n > 2$, then either there is an odd prime divisor p, and $\varphi(n)$ is divisible by $p - 1$, or $n = 2^k$, where $n > 2$ implies $k \geq 2$, and hence $\varphi(n) = 2^{k-1}$. We will prove the following:

Theorem 6 *The density of the numbers that occur as values of the function* φ *is 0. More precisely, for an arbitrary (small) positive number h, there is a bound K such that if x > K, then there are fewer than hx numbers t for which there is an n satisfying* $\varphi(n) = t$.

The proof consists of two parts. On the one hand, we will show that there are only "few" numbers t less than x that can occur as the value of φ on places "having many prime divisors." On the other hand—and this is the crucial part of the proof—there are only few numbers that do not have many prime divisors.

For the purposes of the first part of the proof, we start with a k to be determined later. For numbers n that have at least k odd prime divisors, $\varphi(n)$ is divisible by 2^k; thus for these places, the function can take on at most $x/2^k$ different values. If, on the other hand, n is a number divisible by fewer than k odd divisors, thus

$$n = 2^u p_1^{u_1} \cdots p_j^{u_j}, \quad j < k,$$

and $\varphi(n) \leq x$, then

$$x \geq \varphi(n) \geq n\left(1 - \frac{1}{2}\right)\left(1 - \frac{1}{p_1}\right) \cdots \left(1 - \frac{1}{p_j}\right)$$

$$\geq n\left(1 - \frac{1}{2}\right)\left(1 - \frac{1}{3}\right) \cdots \left(1 - \frac{1}{j+2}\right) = \frac{n}{j+2} \geq \frac{n}{k+1},$$

$$n \leq (k+1)x.$$

In what follows, we will denote $(k+1)x$ by y. We will estimate the number of integers of the form above that are not greater than y (we will not worry about whether or not φ actually takes on distinct values for these integers). Denote by Q the largest of the divisors

$$2^u, p_1^{u_1}, \ldots, p_j^{u_j}.$$

Then

$$n \leq Q^{j+1} \leq Q^k, \qquad Q \geq n^{1/k},$$

and denoting by d be the number obtained by omitting Q from the product, we have

$$d = \frac{n}{Q} \leq n^{(k-1)/k} \leq y^{(k-1)/k}. \tag{7}$$

For such a d, the number of possible prime powers Q is at most y/d (in fact fewer, because Q cannot be smaller than $d^{1/(k-1)}$, since this would contradict Q being the largest prime power divisor of dQ). In this way—denoting by $\pi^*(z)$ the number of prime powers not greater than z—the number of multiples of d that can occur as n is, according to Theorem 5.8 and (7), at most

$$\pi^*\left(\frac{y}{d}\right) \leq \frac{Cy/d}{\log\frac{y}{d}} = \frac{Cy}{d(\log y - \log d)}$$

$$\leq \frac{Cy}{d\left(\log y - \frac{k-1}{k}\log y\right)} = \frac{Cky}{d\log y},$$

where C is an appropriate constant.

The possibilities for d are only those numbers that are divisible by at most k distinct primes. Let us denote the finite sum of such numbers by $\sum^*$, and $y^{(k-1)/k}$ by z; then the number of possible integers n is at most

$$\sum_{d\leq z}^* \frac{Cky}{d\log y} = \frac{Cky}{\log y} \sum_{d\leq z}^* \frac{1}{d}.$$

We can get every term of the sum (and terms larger than z too) if we add one to the sum of the reciprocals of all prime powers not greater than z, and then raise the entire sum to the kth power. For a prime p (not greater than z), denote by r_p the largest exponent for which $p^{r_p} \leq z$. Then

$$\sum_{d\leq z}^* \frac{1}{d} < \left(1 + \sum_{p\leq z}\left(\frac{1}{p} + \frac{1}{p^2} + \cdots + \frac{1}{p^{r_p}}\right)\right)^k$$

$$= \left(1 + \sum_{p\leq z}\frac{1}{p} + \sum_{p\leq z}\frac{1 - \frac{1}{p^{r_p}-1}}{p^2\left(1 - \frac{1}{p}\right)}\right)^k$$

$$< \left(1 + \sum_{p\leq z}\frac{1}{p} + \sum_{p\leq z}\frac{1}{p(p-1)}\right)^k < \left(1 + \sum_{p\leq z}\frac{1}{p} + \sum_{n=2}^{z}\frac{1}{n(n-1)}\right)^k$$

$$= \left(1 + \sum_{p\leq z}\frac{1}{p} + \left(1 - \frac{1}{2}\right) + \left(\frac{1}{2} - \frac{1}{3}\right) + \cdots + \left(\frac{1}{z-1} - \frac{1}{z}\right)\right)^k$$

$$< \left(\sum_{p\leq z}\frac{1}{p} + 2\right)^k.$$

22*. In connection with Theorem 5.4, we mentioned without proof that the difference between the sum in parentheses above and $\log \log z$ is bounded. Here it will be enough to use the following weaker bound, which holds for some appropriate constant c:

$$\sum_{p \leq z} \frac{1}{p} < c \log \log z,$$

and this follows from Theorem 5.7. According to this theorem, if p_n is the nth prime (thus $n = \pi(p_n)$), then for an appropriate constant c_1,

$$n = \pi(p_n) < \frac{c_1 p_n}{\log p_n}, \quad \text{and thus} \quad p_n > \frac{1}{c_1} n \log p_n > \frac{1}{c_1} n \log n.$$

Thus, using twice the inequality $-x \geq \log(1 - x)$, valid for $x < 1$ (see Fact 5), it follows for $n \geq 3$ that

$$\frac{1}{p_n} < \frac{c_1}{n \log n} = \frac{-c_1 \left(\frac{-1}{n}\right)}{\log n} \leq \frac{-c_1 \log \left(1 - \frac{1}{n}\right)}{\log n}$$

$$= -c_1 \left(-\frac{\log \frac{n}{n-1}}{\log n} \right) \leq -c_1 \log \left(1 - \frac{\log \frac{n}{n-1}}{\log n} \right)$$

$$= c_1 (\log \log n - \log \log(n - 1)),$$

since for $n > 2$ we have $1/n < 1$, and $\log(n/(n - 1))/\log n < 1$. Applying this we obtain

$$\sum_{p \leq z} \frac{1}{p} = \frac{1}{2} + \frac{1}{3} + \sum_{n=3}^{\pi(z)} \frac{1}{p_n} < 1 + c_1 \sum_{n=3}^{\pi(z)} (\log \log n - \log \log(n - 1))$$

$$= 1 + c_1 (\log \log(\pi(z)) - \log \log 2) < 1 + c_1 \left(\log \log \frac{c_1 z}{\log z} + \log \frac{1}{\log 2} \right)$$

$$< c_1 \log \log z + 1 + c_1 \log \frac{1}{\log 2} < c_2 \log \log z,$$

where c_2 is an appropriate constant, for z large enough (e.g., $z > e^{c_1}$). Thus for an appropriate constant c_3,

$$\sum_{d \leq z} {}^{*} \frac{1}{d} < (c_2 \log \log z + 2)^k < (c_3 \log \log z)^k.$$

Using this, we get that the number of integers not greater than $y = (k+1)x$, and not divisible by more than k distinct odd primes, since we have $z = y^{(k-1)/k} = ((k + 1)x)^{(k-1)/k}$, is at most

$$\frac{Cky(c_3 \log \log z)^k}{\log x} < \frac{Ck(k + 1)x(c_3 \log \log y)^k}{\log x}$$

$$= \frac{Ck(k + 1)x (c_3 \log(\log x + \log(k + 1)))^k}{\log x}.$$

With this we can get the following upper bound on the number of integers not greater than x that can occur as a value of φ:

$$x\left(\frac{1}{2^k} + \frac{Ck(k+1)\,(c_3\log(\log x + \log(k+1)))^k}{\log x}\right).$$

Choose k so that $1/2^k < h/2$. If we fix k in this way, then the numerator of the second fraction is smaller than $c_4(\log\log x)^k$, for x large enough and c_4 an appropriate constant. It then follows from Fact 7 that for a fixed k, $c_4(\log\log x)^k/\log x$ is arbitrarily small for x large enough, for instance smaller than $h/2$. Thus if x is larger than this bound, then the number of values of the function not greater than x is smaller than hx, which is what we wanted to prove. This proof is due to S. PILLAI.

23. If we denote by $\pi_r(x)$ the number of integers that have r distinct prime divisors, then the proof above gives an upper bound on the sum $\sum_{r\leq k}\pi_r(x)$. For this function $\pi_r(x)$, LANDAU showed that the quotient

$$\frac{\pi_r(x)}{x(\log\log x)^{r-1}\Big/(r-1)!\log x}$$

is arbitrarily close to 1, for x large enough.

For the function σ, we can also show that the sequence of numbers that occur as a value of $\sigma(n)$ has density 0. If, on the other hand, we consider those integers for which our functions take on values less than x, then there exist constants C_1 and C_2 such that the number of integers n satisfying $\varphi(n)\leq x$ is smaller than C_1x and the number of integers n satisfying $\sigma(n)\leq x$ is smaller than C_2x.[15]

ERDŐS also showed the following: Denoting by $A(x)$ the number of integers t not greater than x for which the equation

$$\varphi(n) = t$$

has a solution, for every positive integer k and arbitrary (small) positive h there is a bound N such that for all x greater than N, the following inequalities hold:[16]

$$\frac{x(\log\log x)^k}{\log x} < A(x) < \frac{x}{(\log x)^{1-h}}.$$

It seems to be a very difficult question to determine the asymptotic value of $A(x)$, or even its order of magnitude.

The situation changes significantly if we ask instead for solutions to $n - \varphi(n)$ rather than $\varphi(n)$. For the product of two primes p and q, we have

[15] P. ERDŐS: *Bull. Amer. Math. Soc.* **51** (1945), pp. 540–544.
[16] P. ERDŐS: *Quarterly Journ. of Math.* **6** (1935), pp. 205–213.

$$pq - \varphi(pq) = p + q - 1.$$

Thus, if it is true (and it is most likely to be true if Goldbach's conjecture is true) that every even integer greater than 6 can be written as the sum of two distinct primes, then every odd integer occurs as the value of $n - \varphi(n)$. For the first few integers, $1 = p - \varphi(p)$ for any prime, $2 = 4 - \varphi(4), 3 = 9 - \varphi(9), 4 = 6 - \varphi(6), 5 = 25 - \varphi(25)$, etc. 10 is the first number that cannot be written as $n - \varphi(n)$. It is a recent result that there are infinitely many even integers that cannot be written as $n - \varphi(n)$.[17]

Using a result of L. G. SCHNIRELMAN, that the sequence of numbers of the form $p + q$, where p and q are prime numbers, has positive density,[18] we also know that the sequence of numbers of the form $n - \varphi(n)$ has positive lower density.

In the equation $m = n - \varphi(n), n$ is generally considerably larger than m, as in the case $n = pq$ that we already saw. We have seen that although a large portion of numbers can be written in the desired form, it is still not impossible that if for a given x we consider all numbers of the form $n - \varphi(n)$, for n smaller than x, then their number divided by x may be arbitrarily small, for x large enough.

We saw that the values of the function φ occur rather infrequently among the natural numbers (Theorem 6). Thus there must be certain numbers that the function takes on often. ERDŐS proved the existence of a constant c such that for infinitely many n there are more than n^c solutions to the equation $\varphi(t) = n$. It is quite probable that this is true for every c less than 1, but if this is indeed true, its proof is very likely to be unusually difficult.

The conjecture of CARMICHEL is still undecided, that every value of the function φ is taken at least twice. If, on the other hand, for some k, the equation $\varphi(t) = n$ has k different solutions, then there are infinitely many such values of n.[19] This is also true for $\sigma(n)$ in place of $\varphi(n)$.

24. A very interesting property of the function φ is the following.

Theorem 7 *Starting from a number n_0, we want to build a sequence $n_0, n_1, n_2, \ldots$, for which $\varphi(n_i) = n_{i-1}$, for $i = 1, 2, \ldots$. Then this is possible only if n_0 is of the form $2^j 3^k$, and the same holds for the other n_i's as well.*

We saw that $\varphi(m)$ is even whenever $m > 2$. It is also clear that if $n_0 = 1$, then $n_i = 1$ (or $n_i = 2^i$) $(i = 1, 2, \ldots)$ is an infinite sequence of the desired type; choosing $n_0 = 2^j 3^k$ $(j \geq 1)$, then $n_i = 2^j 3^{k+i}$ is again an example of a sequence of the desired type. Thus the condition is certainly sufficient.

[17] J. BROWKIN, A. SCHINZEL: *Colloquium Math.* **68** (1995), pp. 55–58.

[18] See, for example, E. LANDAU: *Über einige neuere Fortschritte der additiven Zahlentheorie*, Cambridge Tract No. 35. 1937.

[19] P. ERDŐS: *Acta Arithmetica* **4** (1958), pp. 10–17.

We will prove the necessity of the condition in an indirect way. We first note that if there are arbitrarily large elements of the form $2^j 3^k$, then every element is of this form, because if one element is 1, then every element preceding it is also 1, and if $n_i = 2^j 3^k$, where $j \geq 1$, then

$$n_{i-1} = \begin{cases} 2^j \cdot 3^{k-1} & \text{if } k \geq 1, \\ 2^{j-1} & \text{if } k = 0. \end{cases}$$

Because of our indirect assumption that there are elements not of this form, then from a certain point on, all elements share this property. If our sequence has elements of the form $2^j 3^k$ at the beginning, we leave those off, and the resulting sequence still satisfies the required condition, and further, every element is even. If

$$n_{i+1} = 2^j p_1^{j_1} p_2^{j_2} \cdots p_r^{j_r},$$

where $2 < p_1 < p_2 < \cdots < p_r$ are prime, $j \geq 1, p_r > 3, j_s \geq 1 \, (s = 1, 2, \ldots, r)$, then

$$n_i = \varphi(n_{i+1}) = 2^{j-1} p_1^{j_1-1} p_2^{j_2-1} \cdots p_r^{j_r-1} (p_1 - 1)(p_2 - 1) \cdots (p_r - 1),$$

and this is divisible by 2 to a power that is at least $j + r - 1$. Thus, n_{i+1} is divisible by a lower power of 2 than n_i whenever n_{i+1} is divisible by more than one odd prime, and also in the case where it is divisible by only one odd prime and this prime is of the form $4u + 1$. The exponent of 2 cannot decrease infinitely often; thus starting from some point on, every term must be divisible by the same power of 2.

We can restrict ourselves to the sequence starting from this point on. In this sequence every element is of the form $n_i = 2^j p_i^{j_i}$, where p_i is a prime greater than 3 of the form $4u_i + 3, j \geq 1, j_i \geq 1$. Thus

$$n_i = \varphi(n_{i+1}) = 2^j p_{i+1}^{j_{i+1}-1} \frac{p_{i+1} - 1}{2}.$$

Here the last factor is an odd number greater than 1 and less than p_{i+1}. Because n_i has only one odd prime divisor, p_i, it follows that

$$j_{i+1} = 1 \quad \text{and} \quad \frac{p_{i+1} - 1}{2} = p_i; \quad \text{i.e.,} \quad p_{i+1} = 2p_i + 1.$$

This means that starting from an appropriate p_i,

$$p_i, \; p_{i+1} = 2p_i + 1, \; \ldots, \; p_{i+j} = 2p_{i+j-1} + 1, \; \ldots$$

has to be an infinite sequence of primes. It can be shown easily by induction that $p_{i+j} = 2^j p_i + 2^j - 1$. This is, however, not possible. According to Fermat's theorem, when $j = p_i - 1, 2^j - 1$ is divisible by p_i, and hence for this j (and perhaps for smaller indices as well), n_j is a composite number. With this we have arrived at a contradiction, and so the condition of the theorem is necessary as well as sufficient.

25. The proof above also gives a method to find an upper bound $f(x)$ for the length s of a sequence $n_0, n_1, \ldots, n_s$, where n_0 is an arbitrary integer not larger than x and not of the form $2^j 3^k$, and

$$\varphi(n_{i+1}) = n_i \qquad (i = 1, 2, \ldots, s - 1).$$

The bound from our proof is, however, presumably far from the actual order of magnitude.

We saw that if we start with a prime p and consider the sequence $2^j p + 2^j - 1, j = 0, 1, \ldots,$ then the term corresponding to $j = p - 1$ is composite. If we start from the prime 3 or 5, then this is the first composite number in the sequence. The question arises whether this occurs for other primes as well. The conjectured negative response was given by K. Győry.[20] An open problem is whether there are infinitely many primes p for which $4p + 1$ is also prime, though this problem has been considered by many people. If there are indeed infinitely many primes of this form, then this would answer positively a question of ARTIN, who asked whether or not there are infinitely many primes p' for which 2 is a primitive root modulo p'.

Starting from a number n_0, the sequence $n_{i+1} = \varphi(n_i)$, $i = 1, 2, \ldots,$ decreases until $n_i > 1$. Denote by $g(n)$ the smallest index for which $n_g = 1$, starting with $n_0 = n$. SHAPIRO showed that the function g is almost additive, in the sense that if $(n, n') = 1$, then $g(nn')$ can differ from $g(n) + g(n')$, but by at most 1.[21] A result of S. PILLAI shows that[22]

$$\frac{\log n}{\log 3} \le g(n) \le \frac{\log n}{\log 2} + 1.$$

It seems likely that there is a constant c between $1/\log 3$ and $1/\log 2$ such that $g(n')/\log n'$ is arbitrarily close to c, for n' large enough, except for a set of natural numbers having density 0. SHAPIRO considered many further questions in his work.

26. We mentioned before that in contrast to the functions $\sigma(n)$ and $\varphi(n)$, the function $\tau(n)$ takes on every positive integer as a value, and with the exception of 1, it takes on every positive integer infinitely often. This does not mean, however, that the function behaves normally. In the following sections we will see many types of "irregularities" of this function.

The function takes on the value 2 for every prime, and hence infinitely often. We will show that this minimum value can occur so that the values for the two neighboring integers are both arbitrarily large. More precisely:

Theorem 8 *For every number Ω, there is a prime p for which $\tau(p-1) > \Omega$, and $\tau(p+1) > \Omega$.*

---

[20] K. GYŐRY: *Mat. Lapok* **18** (1967), Problem 163, p. 344, Solution *ibid.* **20** (1969), p. 178 (in Hungarian).

[21] H. N. SHAPIRO: *American Math. Monthly* **50** (1943), pp. 18–30.

[22] S. PILLAI: *Bull. Amer. Math. Soc.* (1929), pp. 837–841.

The proof uses Dirichlet's theorem, which we stated in Section 2.19 (and proved only for some special cases). This theorem asserts that every arithmetic progression contains infinitely many primes, as long as the first term and the difference are relatively prime.

We will further use the fact that if a number n is divisible by the odd primes $t_1, t_2, \ldots, t_r$, then

$$\tau(n) \geq 2^r.$$

Based on this, we will choose r so that $2^r > \Omega$, and $2r$ odd primes, say

$$p_1, p_2, \ldots, p_r; \quad q_1, q_2, \ldots, q_r.$$

We then determine all x for which $x - 1$ is divisible by the first r primes, and $x + 1$ is divisible by the second r primes. Thus we are searching for solutions to the simultaneous congruences

$$x \equiv 1 \pmod{p_1 p_2 \cdots p_r},$$
$$x \equiv -1 \pmod{q_1 q_2 \cdots q_r}. \tag{8}$$

Because the coefficient of x is 1 in both congruences, thus relatively prime to the moduli, and the two moduli are relatively prime to each other, since there is no common prime divisor, it follows from Theorem 2.7 that there is a solution, and the solution is a congruence class modulo the product of the different moduli. If x_0 is a solution, then the residue class contains the following arithmetic progression:

$$x_0 + k p_1 p_2 \cdots p_r q_1 q_2 \cdots q_r, \quad k = 0, 1, \ldots .$$

Here x_0 and the difference of the arithmetic progression are relatively prime, because if x_0 were divisible by one of the p or one of the q, then in the first case it would be a divisor of 1, and in the second case a divisor of -1, because 0 is not a solution of (8). This is, however, not possible. Theorem 8 now follows from this, since according to Dirichlet's theorem, there are infinitely many primes in the arithmetic progression, and for every such prime p, we have

$$\tau(p) = 2, \qquad \tau(p-1) \geq 2^r > \Omega, \qquad \tau(p+1) \geq 2^r > \Omega.$$

27. The properties that have been proved so far relating to the number of divisors show that τ can have large jumps for neighboring numbers. On the other hand, it is easy to give certain upper and lower bounds on the order of magnitude of these jumps, even if they are far apart. It is easy to see that $\tau(n)$ takes on values greater than $\log n$ to any arbitrary power.

Theorem 9 *For every positive k, there are infinitely many n for which*

$$\tau(n) > (\log n)^k.$$

It suffices to prove that for every k, there is a positive constant C_k such that

$$\tau(n) > C_k (\log n)^k \tag{9}$$

is satisfied infinitely often. This is because we can rephrase the result with $k+1$ in place of k and consider only those n for which $\log n > 1/C_{k+1}$. There are infinitely many n satisfying this, and for these we have

$$\tau(n) > (C_{k+1} \log n)(\log n)^k > (\log n)^k.$$

Working towards the proof of (9), we note that if for n we chose powers of a fixed m, then $\tau(n)$ is essentially given by the exponent, which is $\log n / \log m$, where m is fixed.

For a given k, let us take k distinct primes $p_1, p_2, \ldots, p_k$ (for example the first k primes), and let $n = (p_1 p_2 \cdots p_k)^r$. Then

$$r = \frac{\log n}{\log(p_1 p_2 \ldots p_k)},$$

and hence

$$\tau(n) = (r+1)^k > r^k = \frac{1}{\log(p_1 p_2 \cdots p_k)^k}(\log n)^k.$$

Here $C_k = 1/\log(p_1 p_2 \cdots p_k)^k$ does not depend on r. For this C_k, (9) is satisfied for all n considered, and we have proved Theorem 9.

28. From the opposite side, $\tau(n)$ is smaller than n to any power, as long as n is large enough.

Theorem 10 *For any positive h, there is a constant K_h for which*

$$\tau(n) < n^h$$

whenever $n > K_h$.

Here, in analogy to the proof of Theorem 9, it is enough to show that there is a constant D_h depending only on h such that

$$\tau(n) < D_h n^h \tag{10}$$

is satisfied for all n above a certain value. If this is true, then we consider all n satisfying the statement for $h/2$ in place of h that in addition satisfy

$$D_{h/2} < n^{h/2}, \quad \text{and thus} \quad n > K_n = D_{h/2}^{2/h}.$$

For these,

$$\tau(n) < D_{h/2}n^{h/2} < n^h.$$

We note that for $h = \frac{1}{2}$, the validity of the claim is obvious. We arrange the divisors of n in pairs (d, d'), so that $dd' = n$, and $d \leq d'$. (We call these pairs *conjugate divisors* of n.) Then $d \leq \sqrt{n}$, and counting roughly, there are at most $\sqrt{n}$ divisor couples, thus at most $2\sqrt{n}$ divisors, and the claim holds letting $D_{1/2}$ equal 2.

To prove (10), we will investigate the quotient $\tau(n)/n^h$. If the canonical decomposition of n is

$$n = p_1^{k_1}p_2^{k_2} \cdots p_r^{k_r},$$

then

$$\frac{\tau(n)}{n^h} = \frac{k_1 + 1}{p_1^{k_1 h}} \cdot \frac{k_2 + 1}{p_2^{k_2 h}} \cdots \frac{k_r + 1}{p_r^{k_r h}}.$$

We will see that the factors on the right-hand side can be larger than 1 for only a finite set of primes, and in fact, this can happen only if the exponent is below some bound. We will first establish the bound on the exponent. From the inequality

$$\frac{k+1}{p^{hk}} > 1 \tag{11}$$

it follows that

$$\log(k + 1) > hk \log p > hk \log 2, \quad \text{i.e.,} \quad \frac{\log(k+1)}{k} > h \log 2.$$

As k becomes large, it follows from Fact 7 that the left-hand side (of the inequality on the right) becomes arbitrarily small. Thus the inequality can hold only for $k + 1$ below a certain bound C_h (depending on h).

From (11), it follows for p using Fact 5 that

$$\log p < \frac{1}{h}\frac{\log(k+1)}{k} < \frac{1}{h}.$$

This shows that only finitely many primes can satisfy the inequality. These two observations together show that only finitely many prime powers can satisfy (11). The number of these (which depends only on h) we denote by m_h, and in the canonical decomposition of n we denote the product of these by n' and the product of the rest by n''. Thus $(n', n'') = 1$, and hence $\tau(n) = \tau(n')\tau(n'')$. Here n' has at most m_k prime power factors, and the exponent of every one satisfies $k + 1 < C_h$; thus

$$\tau(n') < C_h^{m_h}.$$

On the other hand, for the prime power factors of n'',

$$\frac{k+1}{p^{hk}} \leq 1, \quad \text{and hence} \quad k + 1 \leq p^{hk}.$$

Hence

$$\tau(n'') \leq n''^h \leq n^h,$$

and thus

$$\frac{\tau(n)}{n^h} < \frac{\tau(n')\tau(n'')}{n^h} \leq C_h^{m_h},$$

which depends only on h. With this we have proved (10) and with it Theorem 10.

Theorem 10 can be sharpened. WIGERT proved that for every c larger than $\log 2$, there is a bound such that for all n larger than that bound,[23]

$$\tau(n) < n^{c/\log\log n}.$$

RAMANUJAN gave a very simple proof of this theorem, and generalized it. He obtained a good asymptotic estimation for the maximal order of magnitude of the function in connection with *strongly composite* numbers.[24]

29. With the example of the function $\tau(n)$, we have tried to show how the behavior of arithmetic functions can be quite whimsical. On the other hand, GAUSS already noticed that the average values of these functions behave very well. More precisely, the sum of the values of these functions for all natural numbers up to a bound x can be approximated very well by simple functions.

As an example, let us look at $\tau(n)$ and consider the function

$$T(x) = \sum_{n \leq x} \tau(n).$$

This means that for every integer not greater than x, we count each of its divisors once. These are the numbers not greater than x, and such a divisor d is counted once for each of its multiples that is not greater than x. The number of these multiples is $[x/d]$. Summing these for all d, we have

$$T(x) = \sum_{d \leq x} \left[\frac{x}{d}\right].$$

This can be estimated from above and from below by using $[u] > u - 1$ and $[u] \leq u$, respectively, to give

$$T(x) > \sum_{d \leq x} \left(\frac{x}{d} - 1\right) = x\sum_{d \leq x} \frac{1}{d} - [x] \geq x\sum_{d \leq x} \frac{1}{d} - x$$

and

$$T(x) \leq \sum_{d \leq x} \frac{x}{d} = x\sum_{d \leq d} \frac{1}{d}.$$

[23] S. WIGERT: *Arkiv för Math.* 3 (1907), pp. 1–9; S. RAMANUJAN: *Journ. Indian Math. Soc.* **7** (1915), pp. 131–133.

[24] S. RAMANUJAN: *Proc. London Math. Soc.* **14** (1915), pp. 347–409.

Using the inequality in Fact 6, which says that

$$\log(m+1) < \sum_{d=1}^{m} \frac{1}{d} < \log m + 1,$$

we have

$$T(x) > x \log([x]+1) - x > x \log x - x$$

and

$$T(x) < x(1 + \log x) = x \log x + x.$$

This verifies the following:

Theorem 11 *For every x,*

$$\left| \sum_{n \leq x} \tau(n) - x \log x \right| < x.$$

For the average value of $\tau(n)$, we see that replacing x by a natural number n yields

$$\frac{\tau(1) + \tau(2) + \cdots + \tau(n)}{n} = \frac{T(n)}{n} = \log n + r(n),$$

where $r(n)$ depends on n, but its absolute value is smaller than 1.

30. We can obtain a sharper bound for $T(x)$ if for a number n we consider a divisor d together with its conjugate divisor d' (for which $dd' = n$). The smaller of the two divisors (or each divisor in the case that they are equal) is not greater than $\sqrt{n}$. We can add together the smaller of the two conjugate divisors in a way similar to that above, and now use the fact that the number of terms in the sum is at most $[\sqrt{x}]$. We can also use this sharper bound for the larger of the two divisors, and in this way we arrive at the bound

$$T(x) = x \log x + (2c - 1)x + r_1(x),$$

where c is the Euler–Mascheroni constant that we introduced in Section 3.15; it is not difficult to see that the remainder r_1 divided by $\sqrt{x}$ is bounded.

SIERPIŃSKI (1906) proved first, using a method of VORONOI, but then using others as well in four completely different ways, that $r_1(x)/x^{1/3}$ is bounded. So it seemed very probable that $x^{1/3}$ would be the correct order of magnitude of the remainder, until 1932, when VAN DER CORPUT showed that $r_1(x)/x^{0.33}$ is bounded. The best result in this direction is HUXLEY's. According to this, $r_1(x)/x^{\frac{23}{73}+h}$ is bounded for any positive h.[25]

In the other direction, HARDY and LANDAU showed in 1915 that

---

[25] M. N. HUXLEY: *Proc. London Math. Soc.* **66** (1993), pp. 279–301.

$$\frac{r_1(x)}{(x \log x)^{\frac{1}{4}}}$$

does not converge to 0.[26] In spite of many attempts, the order of magnitude of $r_1(x)$ still remains unknown, although it seems likely that it is essentially given by the lower bound.

31. In a similar way, we may investigate the sum of the values of the function $\omega(n)$. Every n not greater than x contributes the number of its distinct prime divisors to the sum. Turning this around, each prime is counted once for each of its multiples up to x, thus $[x/p]$ times. With this we have

$$\sum_{n \leq x} \omega(n) = \sum_{p \leq x} \left[\frac{x}{p}\right] \begin{cases} > \sum_{p \leq x} \left(\frac{x}{p} - 1\right) = x \sum_{p \leq x} \frac{1}{p} - \pi(x) > x \sum_{p \leq x} \frac{1}{p} - x, \\ \leq x \sum_{p \leq x} \frac{1}{p}. \end{cases}$$

The sum here differs at most by a constant from $\log \log x$, using Theorem 5.4 and the remarks made after the theorem (we gave only a lower bound to the difference). In this way,

$$\sum_{n \leq x} \omega(n) = x \log \log x + r_\omega(x),$$

where $|r_\omega(x)|/x$ is bounded. In a similar way, we have

$$\sum_{n \leq x} \Omega(n) = x \log \log x + r_\Omega(x),$$

where $|r_\Omega(x)|/x$ is also bounded.

32. Similar expressions are also known for the related functions $\sigma(n)$ and $\varphi(n)$:

$$\sum_{n \leq x} \sigma(n) = \frac{\pi^2}{6} x^2 + r_\sigma(x),$$

where $|r_\sigma(x)|/(x \log x)$ is bounded;

$$\sum_{n \leq x} \varphi(n) = \frac{3}{\pi^2} x^2 + r_\varphi(x),$$

where $|r_\varphi(x)|/(x \log x)$ is bounded. The later result is due to WALFISZ and was improved a little by KOROBOV, who showed that $|r_\varphi(x)|/x(\log x)^{\frac{5}{7}+h}$ is bounded for every (small) positive h.[27]

---

[26] See E. LANDAU: *Vorlesungen über Zahlentheorie*, S. Hirzel Verlag, Berlin, volume 2, pp. 183–184 and pp.188–189.

[27] N. M. KOROBOV: *Doklady Akad. Nauk SSSR* **119** (1958), pp. 433–434.

In the other direction, it is rather easy to see that the quotient

$$\frac{|r_\sigma(x)|}{x \log \log x}$$

is greater than any bound infinitely often. CHOWLA and S. PILLAI showed that a similar statement is also true for the quotient[28]

$$\frac{|r_\varphi(x)|}{x \log \log \log x}.$$

We note that the average value of one of these functions does not mean that most often it takes on a value near this. For example, the average value of $\tau(n)$ is $\log n$, but it can be shown that the values it takes on most frequently are close to $(\log n)^{\log 2} = (\log n)^{0.69\cdots}$. The bigger average value is a result of the not too frequent, but very large, values of size about $(\log n)^{\log 4}$. It can be shown (and has not been published yet) that for any (small) positive h, if we add together all values of $\tau(n)$ for all n not greater than x and having value either greater than $(\log n)^{\log 4 + h}$, or smaller than $(\log n)^{\log 4 - h}$, and denote this sum by $T^*(n)$, then $T^*(x)/\log x$ is arbitrarily small for x large enough.

HARDY and RAMANUJAN showed that the irregularities in the behavior of the functions $\omega(n)$ and $\Omega(n)$ are not so pronounced; their value is most often close to their average value $\log \log n$. A simple proof of this was given by TURÁN.[29] More precisely, for an arbitrary positive h, those numbers n satisfying

$$|\omega(n) - \log \log n| > h \log \log n$$

have density 0, and the same is true of those n satisfying

$$|\Omega(n) - \log \log n| > h \log \log n.$$

33. We mentioned in Section 15 that classes of multiplicative and additive functions, and even those of completely multiplicative and completely additive functions, still allow for diversity. The question arises as to which additional conditions determine the functions.

Along these lines, we present the following result.

Theorem 12 *If $f(n)$ is an additive function that is nondecreasing, i.e., $f(n) \leq f(n')$ whenever $n \leq n'$, then for an appropriate constant c,*

$$f(n) = c \log n.$$

---

[28] These types of questions are discussed in detail by HARDY and WRIGHT, in their book already cited: *An introduction to the Theory of Numbers*, in chapter XVIII and in chapter XXIII, paragraphs 10–13.

[29] P. TURÁN: *Journ. London Math. Soc.* **9** (1934), pp. 274–276.

Because the function is additive, $f(1) = 0 = c \log 1$ holds for any constant c, and thus for $n = 1$, the theorem is true. For $n > 1$, we can rephrase the claim of the theorem, that if a and b are integers greater than 1, then

$$\frac{f(a)}{\log a} = \frac{f(b)}{\log b}. \tag{12}$$

We will prove this by estimating the two fractions with the help of the fraction $f(m)/\log m$, for a "large" m. Then we will show that the bound we get for the difference of the two sides of (12) can hold for all m only if the difference is 0.

Let m be an arbitrary integer larger than both a and b, and let k be the integer satisfying

$$a^k \le m < a^{k+1}.$$

Then $f(m) \le f(a^{k+1})$. In order to use the fact that the function is additive, increase the argument by a. Because $a^{k+1} + a = a(a^k + 1)$, and $(a, a^j + 1) = 1$, whenever $j \ge 1$, we have

$$\begin{aligned} f(m) &\le f(a^{k+1} + a) = f(a(a^k + 1)) = f(a) + f(a^k + 1) \\ &\le f(a) + f(a(a^{k-1} + 1)) = 2f(a) + f(a^{k-1} + 1) \\ &\le \cdots \le kf(a) + f(a + 1). \end{aligned}$$

In a similar way we can obtain a lower bound for $f(m)$, using that

$$(a, a^j - 1) = 1 \quad \text{whenever} \quad j \ge 1.$$

This gives us

$$\begin{aligned} f(m) &\ge f(a^k) = f(a(a^{k-1} - 1)) = f(a) + f(a^{k-1} - 1) \\ &\ge f(a) + f(a(a^{k-2} - 1)) = 2f(a) + f(a^{k-2} - 1) \\ &\ge \cdots \ge (k-1)f(a) + f(a - 1). \end{aligned}$$

Using these two inequalities, we can estimate $f(m)/\log m$ from both sides. On the one hand,

$$\frac{f(m)}{\log m} \le \frac{kf(a)}{\log m} + \frac{f(a+1)}{\log m} \le \frac{kf(a)}{\log a^k} + \frac{f(a+1)}{\log m} = \frac{f(a)}{\log a} + \frac{f(a+1)}{\log m}.$$

On the other hand,

$$\begin{aligned} \frac{f(m)}{\log m} &\ge \frac{(k-1)f(a)}{\log m} + \frac{f(a-1)}{\log m} = \frac{(k+1)f(a)}{\log m} - \frac{2f(a) - f(a-1)}{\log m} \\ &> \frac{(k+1)f(a)}{(k+1)\log a} - \frac{2f(a) - f(a-1)}{\log m} \\ &= \frac{f(a)}{\log a} - \frac{2f(a) - f(a-1)}{\log m}. \end{aligned}$$

From the two inequalities, we have

$$\frac{f(m) - f(a+1)}{\log m} \le \frac{f(a)}{\log a} < \frac{f(m) + 2f(a) - f(a-1)}{\log m}.$$

The same arguments with b in place of a give, because of $m > b$,

$$\frac{f(m) - f(b+1)}{\log m} \le \frac{f(b)}{\log b} < \frac{f(m) + 2f(b) - f(b-1)}{\log m}.$$

We now multiply all terms of this last inequality by -1, reversing the inequality signs, and then add the corresponding sides to those for $f(a)/\log a$. This gives

$$-\frac{f(a+1) + 2f(b) - f(b-1)}{\log m} < \frac{f(a)}{\log a} - \frac{f(b)}{\log b} \le \frac{2f(a) - f(a-1) + f(b+1)}{\log m}.$$

This pair of inequalities can hold for all sufficiently large m only if

$$\frac{f(a)}{\log a} - \frac{f(b)}{\log b} = 0,$$

because if the difference were negative, then for large enough m, the first expression would be larger than that difference, and if it were positive, for large enough m, the last expression would be smaller than that difference.

Because a and b were arbitrary, denoting the common value of the quotients by c, we have shown for every n that

$$f(n) = c \log n.$$

34. The theorem above is due to Erdős.[30] The proof given above is essentially that of MOSER and LAMBEK.[31] ERDŐS also showed that the conclusion of Theorem 12 remains true if we replace "nondecreasing" by the statement that $|f(n+1) - f(n)|$ should be arbitrarily small for n large enough,[32] a result that has applications to information theory.[33] The important part of this result is that the absolute value of the difference cannot be "too large." With only minor changes to the proof cited, one can show that if $f(n)$ is an additive function and there exists a positive constant K such that $f(n+1) - f(n) > -K/n$ for every n, then $f(n)$ can only be $c \log n$.

Recently these types of results have been sharpened, and the investigations have been extended in different directions, for example to functions

---

[30] P. ERDŐS: *Annals of Math.* **47** (1946), pp. 1–20. See Theorem 11, page 17.

[31] L. MOSER, J. LAMBEK: *Proc. Amer. Math. Soc.* **4** (1953), pp. 544–545.

[32] See the article referred to in footnote 30, Theorem 13, page 18.

[33] See A. RÉNYI: *Mathematica Cluj* **1** (1959), pp. 341–344.

defined on arithmetic progressions, and to complex-valued functions. It can be shown, for example, that $f(n) = c \log n$ if f is completely additive and for every positive $h, |f(n+1) - f(n)| < h$, for n large enough (V. T. Sós); and also if f is additive and for every positive h, $|f(n+1) - f(n)| < h \log n$, for n large enough (E. WIRSING).[34]

Exercises:

31*. Denote by $\sum_2$, respectively $\sum_3$, the sum over composite numbers, respectively the sum for those numbers with at least three prime divisors (not necessarily distinct). Prove that there are constants K_1 and K_2 such that for every x, the following inequalities hold:

$$\sum_{n \leq x} {}_2 \frac{1}{n\,(n - \varphi(n))} < K_1 \quad \text{and} \quad \sum_{n \leq x} {}_3 \frac{1}{(n - \varphi(n))^2} < K_2.$$

32*. Prove that the series

$$\sum_{n=1}^{\infty} \frac{\sigma(n)}{n!}$$

represents an irrational number.

Note. ERDŐS and KAC have shown that $\sum_{n=1}^{\infty} \sigma_2(n)/n!$ represents an irrational number, and I. Z. RUZSA[35] showed it for $\sigma_3(n)$ in place of $\sigma_2(n)$. It is not known whether the same holds for $\sigma_k(n), k > 3$.

---

[34] These types of questions are dealt with in detail in P. D. T. A. ELLIOT: *Arithmetic Functions and Integer Products*, Springer-Verlag, Berlin (1985). See also K. KOVÁCS: *Acta Math. Hung.* **50** (1987), pp. 123–125, and the references cited there.

[35] I. Z. RUZSA, personal communication.

Hints to the More Difficult Exercises

Chapter 1

20*. Start with the product of the first few primes. Adding numbers greater than 1 (or subtracting) we get many consecutive composite numbers.

36. (b*) Show and use that the distinguished common divisor of two numbers and that of one of the numbers and their difference are the same. (Refer to the Euclidean algorithm in Section 27.)

42*. For the first two parts it is helpful to consider the size of $[-x]$; the third function can be written as $\frac{1}{2} \pm \left(x - [x] - \frac{1}{2}\right)$. Here, the sign can be expressed with the help of $[x]$.

43*. Use the identity $n\binom{2n}{n} = (n+1)\binom{2n}{n-1}$, or compare the powers of the prime divisors of $n+1$ in the canonical decomposition of $n+1$ and $\binom{2n}{n}$.

44*. First solution: Use the canonical decomposition of $n!$ to express the canonical decomposition of the quotient. Second solution: Represent the quotient by Q and use the canonical decomposition of $n!$ to show that $[Q/p^r]$ is 0 or 1, for every prime p and every positive exponent r.

52*. This can be solved in a way similar to the proof of Theorem 16.

65*. (a) The hypothesis $(a, b) = 1$ is necessary. The largest x corresponding to a nonnegative y is of the form $c - tb$ for a positive c. For $t = 0, 1, \ldots, a-1$, these numbers form a complete residue system modulo a; thus one of them is divisible by a. For what values of c will this be negative? Count the numbers of the form $ub, a + u'b, 2a + u''b, \ldots$, respectively, that fall below the bound obtained this way. Pay attention to which numbers can occur as a remainder of the divisions inside the integer-part brackets. (All $n \geq (a-1)(b-1)$ can be represented in this way; $\frac{1}{2}(a-1)(b-1)$ numbers cannot.)

Chapter 2

24*. In the polynomial identity of Section 34, substitute p for x and use the fact that the a_i are divisible by p.

25*. (a) This follows from Theorem 13.

25*. (b) Let r be a value satisfying **(a)**, multiply by r^k, and determine the remainder of the sum, modulo p.

Chapter 3

9*. Dividing the equation $ae^2 + be + c = 0$ through by e yields

$$ae + c\frac{1}{e} + b = b + (a + c)\left(1 + \frac{1}{2!} + \frac{1}{4!} + \cdots\right) + (a - c)\left(\frac{1}{1!} + \frac{1}{3!} + \cdots\right) = 0.$$

If the value of $a + c$ or $a - c$ is 0, then we may proceed in a fashion similar to the proof of the irrationality of e. In the remaining cases, we may write the right-hand side as

$$b + \frac{1}{0!}\left(a + c + \frac{a - c}{1}\right) + \frac{1}{2!}\left(a + c + \frac{a - c}{3}\right) + \frac{1}{4!}\left(a + c + \frac{a - c}{5}\right) + \cdots.$$

If n is large enough, then the term containing $1/(2n)!$ has a constant sign, and the same ideas as above will complete the problem.

10*. If a_j and b_j are the first pair that differ, we may assume that $a_j = 2$ and $b_j = 3$. Now bound A_n from below and B_n from above.

11*. Prove that if the sequence of the a_k's is not periodic, then the numbers that arise by multiplying the constructed number by $a_1, a_1a_2, a_1a_2a_3, \ldots$ have infinitely many distinct fractional parts.

12*. Use the fact that an irreducible polynomial with integer coefficients cannot have multiple zeros. Let $f(x)$ be the desired polynomial, $r = u/v$, and α a zero of f. From $f(r) = f(r) - f(\alpha)$, take out $r - \alpha$ and show that the remaining factor differs by less than a nonzero constant depending on α (and not on r), as long as $|r - \alpha|$ is small.

Chapter 4

2*. Let S be the set of positive-integer-valued periodic functions f having period p for which $f(a) = 1$ and $f(p - a) = 0$, and let $(\mathbf{T}f)(x) = f(x + 1)$.

3*. We may assume that $abcd \neq 0$. Let S be set of the smallest nonnegative residues of the numbers $0, a, 2a, \ldots$, modulo c, and for every element of S let $\mathbf{T}m \equiv m + a \pmod{c}$ $(0 \leq \mathbf{T}m \leq c - 1)$. Then $\mathbf{T}^c$ is the identity mapping, and c and b are also periodic.

4*. We may assume that $1 \leq a\, p - 1$. Choose as S the set of positive-valued p-periodic functions that take on values between 0 and $a - 1$, and for which 0 is mapped to 0. Finally, let $(\mathbf{T}f)(x) = f(x + 1) - f(1)$. Then using the theorem of Problem 1 we can arrive at the desired result.

11*. Let PQ be a minimum lattice distance, and POQ a nonobtuse empty lattice triangle with base PQ. On one side of the line through PQ there are only one or two possibilities for O, and in the latter case, $\angle OPQ$ or $\angle OQP$ is a right angle. Reflect P through the midpoint of OQ and through O, and Q through the midpoint of OP and O. The 6 intersecting triangles form a hexagon with center O.

If there were a nonobtuse empty lattice triangle not congruent to POQ, then from a copy of one of these having vertex O we could construct a hexagon centered at O as above. Show that a line OR of this hexagon would have to pass through the center of OPQ, but this is a contradiction.

13. (b*) Two triangles that share a side of maximum length form a parallelogram. Decrease the perimeter of the two triangles by replacing the diagonal with the shorter one unless two half-squares would arise. After finitely many steps, the longest side of the triangles arising in this way will be $\sqrt{5}$, and such triangles will occur in pairs forming parallelograms.

Starting from a segment of unit length of a side of a rectangle, if a parallelogram is leaning against it, then we take the other side of unit length.

Show that by this procedure, starting from two adjacent sides of the rectangle we can reach at most the parallel side. The statement follows immediately.

14*. From the hypothesis on the angles, the area of PQS must be greater than that of PRS; thus it cannot be empty. The proof follows from this observation. (See also Exercise 3 from the 1988 J. Kürschák Competition. *Középiskolai Mat. Lapok* **39** (1989), pp. 56–60 (in Hungarian).)

16*. Let e be a line passing through O and let P and Q be lattice points in the interior of the circle closest to e, one on each side. Show and use the fact that the point R for which $OPRQ$ is a parallelogram is closer to e than one of either P or Q; thus it is outside the circle, and hence either P or Q is closer to e than OR.

19*. Show that the statement is equivalent to saying that the lattice plane Σ passing through E is the closest lattice plane parallel to the base of the pyramid. Let $EFGH$ be the reflection of the base parallelogram $ABCD$ through the vector AE. If Σ were not the closest lattice plane, then one could label the intersection of the closest plane with the parallelepiped $JKLM$ and that with the pyramid $JNPR$. The former would contain a lattice point, while the latter would not; thus the parallelogram intersecting either NP or RP would contain a lattice point. Projecting this parallelogram through C onto Σ, the projection of the lattice point must be among the lattice points of HE or GF, or FE or GH, respectively. Show that in every case there must be a lattice point on the boundary of $JNPR$ as well.

24*. If the lattice does not have lattice points inside the sphere, then shrink about the origin to bring a pair of opposing lattice points onto the surface of the sphere. By shrinking in an appropriate direction, if necessary, we can bring another pair of opposing lattice points onto the surface of the sphere. The volume of the base parallelepiped does not increase during these operations. Using the results of Exercise 11*, we may achieve that the vertices of a triangle lying in a plane parallel to the plane containing the pairs of opposing lattice points move onto the surface of the sphere (and their mirror images as well) without the volume of the base parallelepiped increasing. The 6 desired points can be chosen among the 10 points. Starting from the plane through the 6 points, repeating the process leads to the smallest parallelepiped lattice that does not have any lattice points in the interior of the sphere except for the origin.

26*. Apply Minkowski's theorem to the (bounded) cylinder of radius $2r$ and height q.

27*. We may give an alternative proof of Theorem 12, similar to the one given, by considering the double cone that intersects the region and satisfies the condition $\left|\left(x^2 + y^2\right) z\right| \leq c$; determine c such that Minkowski's theorem may be applied.

Chapter 5

2*. Show and use that the number between a pair of twin primes must be divisible by 6.

5. (b*) Consider that the number of integers that are square-free and the product of primes not greater than K is not dependent on x.

8*. Consider the solution to the previous problem from the point of view of divisibility by 2 and 3.

11*. Prove and use the fact that if p is a prime, then the prime divisors of $2^p - 1$ are of the form $pk + 1$.

14*. The claim can be proved by induction.

15*. Use the fact that there is a prime between x and $2x$. At least how many elements must there be so that the sum is an integer?

16*. If f is a polynomial of degree at least 1, then it takes on the values -1, 0, 1 at most finitely many times. If $|f(b)| > 1$, then $f(x + b) = f_1(x)$ has constant term $a_0 = f(b)$. (It is possible that b is 0.) Then with the exception of finitely many values of c, $f(ca_0 + b)$ is composite.

Chapter 6

10*. Let $1 = b_1 < c_1 < b_2 < c_2 < \cdots$, and let the sequence be made up of all integers n satisfying $b_i \leq n_i \leq c_i$, for $i = 1, 2, \ldots$. We can arrive at our goal if the b_i's and c_i's grow to the necessary size.

15*. Consider the first k elements of the a_i's, and show that those numbers not divisible by these have asymptotic density, and as k grows, this does not increase. Estimate the number of terms in the sequence from above and below with the help of the densities just determined (in the latter case, including part of the bounded sequence as well).

18*. From among the numbers divisible by only the primes $p_1, \ldots, p_s$, take any infinite subsequence $a_1, a_2, \ldots, a_n, \ldots$. Show that it is possible to choose a subsequence of this such that the exponent of p_1 in the canonical decomposition of the numbers does not decrease; repeating the same for the remaining

p_i's, we arrive at an infinite subsequence in which every element is divisible by the previous one, proving even more than was required.

20*. Partition the elements into sets with the property that every element is larger than the element preceding it (in the given order) and no additional elements can be added. Investigate these sequences and the sequence of their last elements.

21*. Partition the $k\ell + 1$ integers into sets. Let the first set consist of all integers that do not have a divisor among all others. Let the second set be all remaining integers that do not have a divisor among the remaining integers, and so on. Consider the size and number of sets in this partition.

Chapter 7

3*. We may assume that $x > u \geq 0$. Writing the equation as $(x - u)(x + u) = (y - z)(y + z)$, and introducing the parameter $a = (x - u, y - z)$, the problem can be solved by investigating relative primeness and parity.

9*. How many k-element sums can be written with kth powers not greater than k?

12*. From the upper bound (which we will denote by m), remove the largest square, then from the remainder remove again the largest square. Estimate the remainder of this from above with the help of m. Using this upper bound, show that the sum of the two squares cannot be greater than p. (It is clear that it cannot be greater than m.)[36]

17*. Notice that the central number must be relatively prime to the others, and that its square is one more than the product of the others.

18*. (a) Show that the number following the sequence is a perfect square.

18*. (b) Show that one of the factors is relatively prime to the others. It must then be a perfect power, and the product of the remaining factors must also be a perfect power to the same exponent.

19*. The only primes that can be divisors of the l.c.m. of two numbers are those for which at least two multiples occur among the given numbers, thus only the primes less than 16. Investigate the possible ways in which this can occur.

22*. Use the result from the preceding exercise.

23*. Consider the substitutions $(1, 0), (0, 0), (-1, 0)$, and $(0, 1), (0, 0), (0, -1)$. By examining these, it follows that in one of the expressions a and c are both integers, and in the other b and d are both integers. If they are not integers in the same expression, then substituting $(1, 1)$ will lead to the solution.

25*. This can be proved indirectly, using Theorem 1.15. Choose a solution satisfying the hypothesis with the value of z as small as possible. Notice

---

[36] Cf. J. SURÁNYI: *Matematikai Versenytételek III*, Tankönyvkiadó, Budapest (1992), pp. 126–130 (in Hungarian).

that the Pythagorean triple x^2, y^2, z must be primitive. Starting from the canonical decomposition for kth powers which follows from Section 1.19, it is possible to rewrite the decomposition, arriving at a new solution with a smaller z. (Pay attention to which parameters can be even and which ones cannot.)

26*. Apart from the trivial solution $x = y$, if x, y is a solution, we may assume $x < y$ and write $y = tx$ (note that t is rational if both x and y are).

(a) Express the unknowns in terms of t, taking into account that $x \geq 2$.

(b) It is helpful to write $t = 1 + u$. Determine how large the numerator of u can be, given that x is rational.

Chapter 8

4*. Look for n among the powers of two.

11*–12*. Use that the product $\prod_{p \leq x} \left(1 + \frac{1}{p}\right)$ grows beyond all bounds, and respectively that $\prod_{p \leq x} \left(1 - \frac{1}{p}\right)$ diminishes below all positive bounds, for x large enough, and these properties hold, even if we exclude any finite number of factors from either of the products.

30*. Estimate $\varphi(n)$ from above in such a way that we keep only the factor $1 - \frac{1}{p}$. Prove and use the fact that for $k = 2, 3$,

$$\frac{1}{\sqrt[k]{n}} < nk \left(\frac{1}{\sqrt[k]{n-1}} - \frac{1}{\sqrt[k]{n}}\right).$$

31*. Assume that the sum is rational, and let p be a prime larger than the denominator. If we multiply the sum by $p!$, the result must be an integer. This leads to a contradiction starting from the $(p+1)$st term using that $\sigma(n)$ is less than n plus the sum of the first $n/2$ integers.

References

1. Alexanderson, G. L., Pedersen, J.: George Pólya: his life and work, *The Oregon Math. Teacher,* 1985.
2. Avidon, M.: On the distribution of primitive abundant numbers, *Acta Arithmetica,* 1996 **77**, pp. 195–205.
3. Baker, A.: *New Advances in Transcendence Theory,* Cambridge, 1988.
4. Baker, A.: *Transcendental Number Theory,* 3rd ed., Cambridge, 1990,
5. Baker, R. C., Harman, G.: The difference between consecutive primes, *Proc. London Math. Soc.* (3) **72**, 1996, pp. 261–280.
6. Baker, R. C., Harmam, G., Pintz, J.: The difference between consecutive primes II., *Proc. London Math. Soc.* (3) **83**, 2001, pp. 532–562.
7. Balasubramanian, R., Deshuillers, J-M.,Dress, F.: Problème de Waring pour les bicarrés. II. Résultats auxiliaires pour le théorème asymptotique [Waring's problem for biquadrates. II. Auxiliary results for the asymptotic theorem], *Comptes Rendus, Acad. Sci. Paris,* Ser. 1 **303** (1986), pp. 161–163.
8. Balasubramanian, R., Deshuillers, J-M.,Dress, F.: Problème de Waring pour les bicarrés. I. Schéma de la solution [Waring's problem for biquadrates. I. Sketch of the solution], *Comptes Rendus, Acad. Sci. Paris,* Ser. 1 **303** (1986), pp. 85–88.
9. Barnes, E. S., Swinnerton-Dyer, H. P. F.: The inhomogeneous minima of binary quadratic forms I., *Acta Mathematica* **87** (1952), pp. 259–323.
10. Barnes, E. S., Swinnerton-Dyer, H. P. F.: The inhomogeneous minima of binary quadratic forms II., *Acta Mathematica* **88** (1952), pp. 279–316.
11. Barnes, E. S., Swinnerton-Dyer, H. P. F.: The inhomogeneous minima of binary quadratic forms III., *Acta Mathematica* **92** (1954), pp. 199–234.
12. Behrend, F.: Über numeri abundantes, *Sitzungsberichte der Preussischen Akademie der Wissenschaften* **22** (1932), pp. 322–328.
13. Behrend, F.: Über numeri abundantes II., *Sitzungsberichte der Preussischen Akademie der Wissenschaften* **6** (1933), pp. 280–293.
14. Behrend, F.: On sequences of numbers not divisible one by another, *Journal London Math. Soc.* **10** (1935), pp. 42–44.
15. Besicovitch, A. S.: On the density of certain sequences of integers, *Math. Annalen* **110** (1934), pp. 336–341.
16. Beukers, F.: A note on the irrationality of $\zeta(2)$ and $\zeta(3)$, *Bull. London Math. Soc.* **11** No. 3 (1979), pp. 268–272.
17. Birch, B. J.: A grid with no split parallelepiped, *Proc. Cambridge Phil. Soc.* **53** (1956), p. 536.
18. Blichfeldt, H. F.: A new principle in the geometry of numbers, with some applications, *Transactions Amer. Math. Soc.* **15** (1914), pp. 227–235.
19. Browkin, J., Schinzel, A.: On integers not of the form $n - \phi(n)$, *Colloquium Math.* **68** (1995), pp. 55–58.
20. Bugeaud, Y., Győry, K.: Bounds for the solutions of Thue–Mahler equations and norm form equations, *Acta Arithmetica* **74** (1996), pp. 273–292.

21. Bugeaud, Y., Győry, K.: Bounds for the solutions of unit equations, *Acta Arithmetica* **74** (1996), pp. 67–80.
22. Burgess, D.: The distribution of quadratic residues and non-residues, *Mathematika* **4** (1957), pp. 106–112.
23. Cassels, J. W. S.: *An Introduction to the Geometry of Numbers,* Springer, 1959.
24. Chaundy, T. W.: The arithmetic minima of positive quadratic forms I., *Quarterly Journal of Mathematics,* Oxford Ser. **17** (67) (1946), pp. 166–192.
25. Choi, S. L. G.: Covering the set of integers by congruence classes of distinct moduli, *Mathematics of Computation* **25** (1971) pp. 885–895.
26. Davenport, H., Mahler, K.: Simultaneous Diophantine approximation, *Duke Math. Journ.* **13** (1946), pp. 105–111.
27. Davenport, H.: Über numeri abundantes, *Sitzungsberichte der Preussischen Akademie der Wissenschaften* **27** (1933), pp. 830–837.
28. Davenport, H.: Note on the product of three homogeneous linear forms, *Journal London Math. Soc.* **16** (1941), pp. 98–101.
29. Davenport, H.: The geometry of numbers, *Math. Gazette* **31** (1947), pp. 206–210.
30. Davenport, H.: On the product of three homogeneous linear forms II., *Proc. London Math. Soc.* **44** (1938), pp. 412–431.
31. Dedekind, R.: *Stetigkeit und irrationale Zahlen,* Friedrich Vieweg & Sohn, Braunschweig, 1872.
32. Delone, B. N.: An algorithm for the "divided cells" of a lattice (Russian), *Izvestiya Akad. Nauk SSSR.* Ser. Mat. **11** (1947), pp. 505–538 (in Russian).
33. Delone, B. N.: Algorithmus der zerteilten Parallelogramme, *Sowjetwissenschaft* **2** (1948), pp. 178–210.
34. Deshouillers, J.-M.: Problème de Waring pour les bicarrés [Waring's problem for biquadrates], *Seminar on Number Th.,* 1984–1985. Exp. No. **14** 47 pp., Univ. Bordeaux I, Talance, 1985.
35. Dickson, L. E.: Finiteness of the odd perfect and primitive abundant numbers with n distinct prime factors, *Amer. Journ. of Math.* **35** (1913), pp. 413–422.
36. Elkies, N. D.: On $A^4 + B^4 + C^4 = D^4$, *Math. Comp.* **51** (1988), pp. 825–835.
37. Elliot, P. D. T. A.: *Arithmetic Functions and Integer Products,* Springer-Verlag, Berlin, 1985.
38. Elliot, P. D. T. A.: *Probabilistic Number Theory I–II,* Springer-Verlag, Berlin, 1979.
39. Erdős, P., Mahler, K.: On the number of integers which can be represented by a binary form, *Journal London Math. Soc.* **13** (1938), pp. 134–139.
40. Erdős, P., Sárközy, A., Sós, V. T.: On sum sets of Sidon sets I., *Journal of Number Theory* **47** (1994), pp. 329–347.
41. Erdős, P., Selfridge, J. L.: The product of consecutive integers is never a power, *Illinois Journ. of Math.* **19** (1975), pp. 292–301.
42. Erdős, P., Stewart, C. L., Tijdeman, R.: Some Diophantine equations with many solutions, *Compositio Math.* **66** (1988), pp. 37–56.
43. Erdős, P., Szemerédi, E.: On a problem of P. Erdős and S. Stein, *Acta Arithmetica* **15** (1968), pp. 85–90.
44. Erdős, P., Turán, P.: On a problem in the elementary theory of numbers, *American Math. Monthly* **41** (1934), pp. 608–611.
45. Erdős, P., Turán, P.: On some new questions on the distribution of prime numbers, *Bulletin Amer. Math. Soc.* **54** (1948), pp. 371–378.
46. Erdős, P., Turán, P.: On a problem of Sidon in additive number theory, and on some related problems, *Journal London Math. Soc.* **16** (1941), pp. 212–215.
47. Erdős, P.: Some remarks on Euler's φ function, *Acta Arithmetica* **4** (1958), pp. 10–17.

48. Erdős, P.: Remarks on number theory. I. On primitive α-abundant numbers, *Acta Arithmetica* **5** (1959), pp. 25–33.
49. Erdős, P.: Beweis eines Satzes von Tschebyschef, *Acta Litt. Univ. Sci., Szeged, Sect. Math.* **5** (1932), pp. 194–198.
50. Erdős, P.: On the distribution function of additive functions, *Annals of Math.* **47** (1946), pp. 1–20.
51. Erdős, P.: Some remarks on Euler's ϕ-function and some related problems, *Bull. Amer. Math. Soc.* **51** (1945), pp. 540–544.
52. Erdős, P.: On the difference of consecutive primes, *Bull. Amer. Math. Soc.* **54** (1948), pp. 885–889.
53. Erdős, P.: Note on sequences of integers no one of which is divisible by any other, *Journal London Math. Soc.* **10** (1935), pp. 126–128.
54. Erdős, P.: On the density of the abundant numbers, *Journal London Math. Soc.* **9** (1934), pp. 278–282.
55. Erdős, P.: A theorem of Sylvester and Schur, *Journal London Math. Soc.* **9** (1934), pp. 282–288.
56. Erdős, P.: Remarks on two problems (Hungarian), *Mat. Lapok* **11** (1960), pp. 26–33.
57. Erdős, P.: Solution to problem 97 (Hungarian), *Mat. Lapok* **11** (1960), pp. 152–154.
58. Erdős, P.: Über die Reihe $\sum \frac{1}{p}$, *Oberdruk uit Math. B.* **7** (1938), pp. 1–2.
59. Erdős, P.: Problem 97, *Mat. Lapok* **8** (1957), p. 142.
60. Erdős, P.: On additive arithmetical functions and applications of probability to Number Theory, *Proc. Internat. Cong. Mat.*, Amsterdam, Groningen, E. P. Noordhoff, 1954, Vol. III, pp. 13–19.
61. Erdős, P.: On a new method in elementary number theory which leads to an elementary proof of the prime number theorem, *Proceedings Nat. Acad. Sci.* **35** (1949), pp. 374–384.
62. Erdős, P.: Problems and results on the differences of consecutive primes, *Publicationes Math. Debrecen* **1** (1949–1950), pp. 33–37.
63. Erdős, P.: On amicable numbers, *Publicationes Math. Debrecen* **4** (1955), pp. 108–111.
64. Erdős, P.: On the difference of consecutive primes, *Quarterly Journ. of Math.*, Oxford Ser. **6** (1935), pp. 124–128.
65. Erdős, P.: On the normal number of prime factors of $p - 1$ and some related problems concerning Euler's ϕ-function, *Quarterly Journ. of Math.* **6** (1935), pp. 205–213.
66. Erdős, P.: On integers of the form $2^k + p$ and some related problems, *Summa Brasiliensis Math. II,* (1950), pp. 113–123.
67. Erhart, E.: Une généralisation du théorème de Minkowski, *Comptes Rendus Acad. Sci. Paris* **240** (1955), pp. 483–485.
68. Euclid. *Elements, On Thomas L. Heath, New York, Dover Publications, 1956.*
69. Furstenberg, H., Katznelson, Y., Ornstein, D. S.: The ergodic theoretical proof of Szemerédi's theorem, *Proc. Symp. Pure Math.* (The mathematical heritage of Henri Poincaré) (**39**, Part 2) 1980, pp. 217–242.
70. Furstenberg, H., Katznelson, Y: A density version of the Hales–Jewett theorem, *Journ. Analyse Mat.* **57** (1991), pp. 64–119.
71. Furstenberg, H.: Ergodic behavior of diagonal measures and a theorem of Szemerédi on arithmetic progressions, *Journ. Analyse Mat.* **31** (1977), pp. 204–256.
72. Gauss, C. F.: *Disquisitiones Arithmeticae,* 1801.
73. Grosswald, E.: *Representations of Integers as Sums of Squares,* Springer, New York, 1985.
74. Gruber, P., Lekkerkerker, C. G.: *Geometry of Numbers,* North Holland, 1987.

75. Győry, K., Sárközy, A., Stewart, C. L.: On the number of prime factors of integers of the form $ab + 1$, *Acta Arithmetica* **74** (1996), pp. 365–385.

76. Győry, K., Stewart, C. L., Tijdeman, R.: On prime factors of sums of integers I., *Compositio Math.* **59** (1986), pp. 81–89.

77. Győry, K.: On the Diophantine equation $\binom{n}{k} = x^l$, *Acta Arithmetica* **80** (1997) pp. 285–295.

78. Győry, K.: On the Diophantine equation $x^p + y^p = cz^p$ (Hungarian), *Mat. Lapok* **18** (1967), Problem 163, p. 344, Solution ibid. **20** (1969), p. 178.

79. Győry, K.: On the numbers of families of solutions of systems of decomposable form equations, *Publ. Math. Debrecen* **42** (1993), pp. 65–101.

80. Hadamard, J.: Sur la distribution des zéros de la fonction $\zeta(s)$ et ses conséquences arithmétiques, *Bulletin Soc. Math. France* **24** (1896), pp. 199–220.

81. Hajós, Gy.: Ein neuer Beweis eines Satzes von Minkowski, *Acta Sci. Math. Szeged* **6** (1934), pp. 224–225.

82. Hardy, G. H., Wright, E. M.: *An Introduction to the Theory of Numbers*, Oxford, 1960.

83. Härtig, K., Surányi, J.: Combinatorial and geometrical investigations in elementary number theory, *Periodica Math. Hung.* **6** (1975), pp. 235–240.

84. Hegyvári, N.: On some irrational decimal fractions, *The American Mathematical Monthly* **100** (1993), pp. 779–780.

85. Hoffmann, Gy., Surányi, L.: An exposition of the first arithmetic proof of the prime number theorem (Hungarian), *Matematikai Lapok* **23** (1972), pp. 31–51.

86. Hornfeck, B., Wirsing, E.: Über die Häufigkeit vollkommener Zahlen, *Math. Annalen* **133** (1957), pp. 431–438.

87. Huxley, M. N.: Exponential sums and lattice points II., *Proc. London Math. Soc.* **66** (1993), pp. 279–301.

88. Kac, M.: Probability methods in some problems of analysis and number theory, *Bull. Amer. Math. Soc.* **55** (1949), pp. 641–665.

89. Korkine, A., Zolotareff, G.: Sur les formes quadratiques, *Math. Annalen* **6** (1873), pp. 366–389.

90. Korobov, N. M.: New number-theoretic estimates (Russian), *Doklady Akad. Nauk SSSR* **119** (1958), pp. 433–434.

91. Kovács, K.: On the characterization of additive functions on residue classes, *Acta Math. Hung.* **50** (1987), pp. 123–125.

92. Kubilius, I. P.: Probabilistic methods in the theory of numbers, *Amer. Math. Transl.* (2) **19** (1962), pp. 47–85.

93. Kubilius, I. P.: Probabilistic methods in number theory (Russian), *Uspekhi Math. Nauk., (N.S.)* **11** (1956), no. 2 (68) pp. 31–66.

94. Landau, E.: *Handbuch der Lehre von der Verteilung der Primzahlen*, 1909, B. G. Teubner, pp. 436–446.

95. Landau, E.: *Vorlesungen über Zahlentheorie*, S. Hirzel, Leipzig, 1927. Vols. I and II.

96. Landau, E.: *Über einige neuere Fortschritte der additiven Zahlentheorie*, Cambridge Tract No. 35. 1937.

97. Lindström, B.: An inequality for B_2-sequences, *Journ. Comb. Theory* **6** (1969), pp. 211–212.

98. Mahler, K.: On lattice points in a cylinder, *Quarterly Journal of Math.*, Oxford Ser. **17** (1946), pp. 16–18.

99. Maier, H., Pommerence, C.: Unusually large gaps between consecutive primes, *Transactions of the Amer. Math. Soc.* **322** (1990), pp. 201–237.

100. Maier, H.: Chains of large gaps between consecutive primes, *Advances in Math.* **39** (1981) pp. 257–269.

101. Maier, H. C.: Small differences between prime numbers, *Michigan Math. Journ.* **35** (1988), pp. 323–344.

102. Markoff, A. A.: Sur les formes quadratiques binaires indéfinies, *Math. Annalen* **15** (1879), pp 381–406.

103. *Mat. és Fiz. Lapok* **50** (1943), pp. 182–183, problem 12. (Hungarian).

104. Mills, W. H.: A prime-representing function, *Bulletin Amer. Math. Soc.* **53** (1947), p. 604.

105. Minkowski, H.: *Geometrie der Zahlen*, Leipzig, 1897.

106. Minkowski, H.: *Diophantische Approximation*, Leipzig, 1907.

107. Mordell, L. J.: *Diophantine Equations*, Academic Press, 1969.

108. Moser, L., Lambek, J.: On monotone multiplicative functions, *Proc. Amer. Math. Soc.* **4** (1953), pp. 544–545.

109. Niven, I., Zuckerman, H. S.: *An Introduction to the Theory of Numbers*, Wiley, New York, 1991.

110. Obláth, R.: Note on the binomial coefficients, *Journal London Math. Soc.* **23** (1948), pp. 252–253.

111. Pick, G.: Geometrisches zur Zahlenlehre, *Lotos Prag* (2) **19** (1900), pp. 311–319.

112. Pillai, S.: On a function connected with $\phi(n)$, *Bull. Amer. Math. Soc.*, (1929), pp. 837–841.

113. Pillai, S. S., Chowla, S. D.: On the error terms in some asymptotic formulae in the theory of numbers I., *Journal London Math. Soc.* **5** (1930), pp. 95–101.

114. Rényi, A.: On a theorem of P. Erdős and its application in information theory, *Mathematica Cluj* **1** (1959), pp. 341–344.

115. Rademacher, H.: *Lectures on Elementary Number Theory*, Blaisdell Publ. Co., New York, 1964.

116. Ramanujan, S.: On the number of divisors of a number, *Journ. Indian Math. Soc.* **7** (1915), pp. 131–133.

117. Ramanujan, S.: Highly composite numbers, *Proc. London Math. Soc.* **14** (1915), pp. 347–409.

118. Rankin, R. A.: The difference between consecutive prime numbers, *Journal London Math. Soc.* **13** (1938), pp. 242–247.

119. Rankin, R. A.: The difference between consecutive prime numbers III., *Journal London Math. Soc.* **22** (1947), pp. 226–230.

120. Reyssat, E.: Seminaire Delaunay–Pisot–Poitou 20ème Année, 1978/79, *Théorie des Nombres*, Fasc. 1 No. 6 6pp.

121. Ribenboim, P.: *The Little Book of Big Primes*, Springer-Verlag, Berlin, 1991.

122. Riemann, B.: Über die Anzahl der Primzahlen unter einer gegebenen Grösse, *Monatsberichte d. Berliner Acad. d. Wiss.*, (1859), pp. 671–680.

123. Roth, K. F.: Rational approximations to algebraic numbers, *Mathematica* **2** (1955), pp. 1–120 and p. 168.

124. Ruzsa, I. Z.: Arithmetical functions I., II. (Hungarian), *Mat. Lapok* **27** (1976–1979), pp. 95–143 and pp. 227–283.

125. Sawyer, D. B.: The product of two non-homogeneous linear forms, *Journal London Math. Soc.* **23** (1948), pp. 250–251.

126. Sawyer, W. W.: *What Is Calculus About?* S.M.S.G., Math. Assoc. of America, 1961.

127. Scherrer, W.: Dissertation, Universität Zürich, 1923.

128. Scherrer, W.: Zur Geometrie der Zahlen, *Math. Annalen* **89** (1923), pp. 255–259.

129. Schmidt, W. M.: *Diophantine Approximations and Diophantine Equations*, Springer, 1991.

130. Schoenberg, I. J.: Über die asymptotische Verteilung reeller Zahlen mod 1, *Math. Zeitscrift* **28** (1928), pp. 171–199.

131. Selberg, A.: An elementary proof of the prime-number theorem, *Annals of Math.* **50** (1949), pp. 305–313.

132. Shapiro, H. N.: An arithmetic function arising from the ϕ function, *American Math. Monthly* **50** (1943), pp. 18–30.

133. Shapiro, H. N.: Note on a theorem of Dickson, *Bull. Amer. Math. Soc.* **55** (1949), 450–452.

134. Shorey, T. N., Tijdeman, R.: *Exponential Diophantine Equations*, Cambridge, 1986.

135. Sidon, S.: Ein Satz über trigonometrische Polynome und seine Anwendung in der Theorie der Fourier-Reihen, *Math. Annalen* **106** (1932), pp. 536–539.

136. Sierpiński, W.: Sur une formule donnant tous les nombres premiers, *Comptes Rendus Acad. Sci.*, Paris **235** (1952), pp. 1078–1079.

137. Singh, S.: *Fermat's Enigma*, Walker & Co. New York, 1997.

138. Sós, V. T.: On the theory of Diophantine approximations II. Inhomogeneous problems, *Acta Math. Hung.* **IX** (1958), pp. 229–241.

139. Sós, V. T.: On the theory of Diophantine approximations I., *Acta Math. Hung.* **VIII** (1957), pp. 462–472.

140. Sós, V. T.: On the distribution mod 1 of the sequence $n\alpha$, *Annales Univ. Sci. Eötvös L. Sect. Math.* **1** (1958), pp. 127–134.

141. Surányi, J.: Über zerteilte Parallelogramme, *Acta Sci. Math. Szeged* **22** (1961), pp. 85–90.

142. Surányi, J.: *Matematikai Versenytételek III.*, Tankönyvkiadó, Budapest, (1992).

143. Surányi, J.: Schon die alten Griechen haben es gewußt, In Freud, R. (ed) *Große Augenblicke aus der Geschichte der Mathematik*, Akadémia Kiadó (1990), pp. 9–50.

144. Székely, J. G. (ed): *Contests in Higher Mathematics,* Springer, New York, 1996.

145. Szemerédi, E.: On sets of integers containing no k elements in arithmetic progression. Collection of articles in memory of Yuriĭ Vladimirovič Linnik, *Acta Arithmetica* **27** (1975), pp. 199–245

146. Taylor, L. E.: Letter to the Editor, *The American Mathematical Monthly* **101** (1994), p. 174.

147. Tijdeman, R.: On the equation of Catalan, *Acta Arith.* **29** (1976), pp. 197–209.

148. Turán, P.: Az egész számok bizonyos számsorozatairól, *Középisk. Mat. Lapok,* new series **8** (1954), pp. 33–41.

149. Turán, P.: On a theorem of Hardy and Ramanujan, *Journ. London Math. Soc.* **9** (1934), pp. 274–276.

150. Vallée-Poussin, Ch. de la: Recherches analytiques sur la théorie des nombres premiers I., *Annales Soc. Sci. Bruxelles* **20** (1896), pp. 183–256.

151. Vallée-Poussin, Ch. de la: Recherches analytiques sur la théorie des nombres premiers II., *Annales Soc. Sci. Bruxelles* **20** (1896), pp. 281–297.

152. Van der Poorten, A.: A proof that Euler missed. . . Apéry's proof of the irrationality of $\zeta(3)$. An informal report, *Math. Intelligencer* (**1**), 1979, pp. 195–203.

153. Van der Waerden, B. L.: Beweis einer Baudetschen Vermutung, *Nieuw Arch. Wisk.* **15** (1927), pp. 212–216.

154. N. B. VASIL'EV, A. A. EGOROV: *Zadachi vsesoyuznikh matematicheskikh Olimpiad (Problems of the All-Soviet-Union Mathematical Olympiads)*, NAUKA, MOSCOW, 1988 (IN RUSSIAN).

155. VAUGHAN, R. C.: A NEW ITERATIVE METHOD IN WARING'S PROBLEM, *Acta Math.* **162** (1989), PP. 1–71.

156. VOROBEV, N. N.: *The Fibonacci Numbers,* HEATH, BOSTON, 1963.

157. WARING, E.: *Meditationes Arithmeticae,* CAMBRIDGE, 1770, THEOR. XLVII.

158. WIGERT, S.: SUR L'ORDRE DE GRANDEUR DU NOMBRE DES DIVISEURS D'UN ENTIER, *Arkiv för Math.* **3** (1907), PP. 1–9.

159. WOOLEY, T. D.: LARGE IMPROVEMENTS IN WARING'S PROBLEM, *Annals of Math.* **135** (1992), PP. 131–164.

160. THE PRIME PAGE: http://www.utm.edu/research/primes/

Index

Undergraduate Texts in Mathematics

(continued from page ii)

Frazier: An Introduction to Wavelets Through Linear Algebra

Gamelin: Complex Analysis.

Gordon: Discrete Probability.

Hairer/Wanner: Analysis by Its History. *Readings in Mathematics.*

Halmos: Finite-Dimensional Vector Spaces. Second edition.

Halmos: Naive Set Theory.

Hämmerlin/Hoffmann: Numerical Mathematics. *Readings in Mathematics.*

Harris/Hirst/Mossinghoff: Combinatorics and Graph Theory.

Hartshorne: Geometry: Euclid and Beyond.

Hijab: Introduction to Calculus and Classical Analysis.

Hilton/Holton/Pedersen: Mathematical Reflections: In a Room with Many Mirrors.

Hilton/Holton/Pedersen: Mathematical Vistas: From a Room with Many Windows.

Iooss/Joseph: Elementary Stability and Bifurcation Theory. Second edition.

Irving: Integers, Polynomials, and Rings: A Course in Algebra

Isaac: The Pleasures of Probability. *Readings in Mathematics.*

James: Topological and Uniform Spaces.

Jänich: Linear Algebra.

Jänich: Topology.

Jänich: Vector Analysis.

Kemeny/Snell: Finite Markov Chains.

Kinsey: Topology of Surfaces.

Klambauer: Aspects of Calculus.

Lang: A First Course in Calculus. Fifth edition.

Lang: Calculus of Several Variables. Third edition.

Lang: Introduction to Linear Algebra. Second edition.

Lang: Linear Algebra. Third edition.

Lang: Short Calculus: The Original Edition of "A First Course in Calculus."

Lang: Undergraduate Algebra. Second edition.

Lang: Undergraduate Algebra. Third edition

Lang: Undergraduate Analysis.

Laubenbacher/Pengelley: Mathematical Expeditions.

Lax/Burstein/Lax: Calculus with Applications and Computing. Volume 1.

LeCuyer: College Mathematics with APL.

Lidl/Pilz: Applied Abstract Algebra. Second edition.

Logan: Applied Partial Differential Equations, Second edition.

Lovász/Pelikán/Vesztergombi: Discrete Mathematics.

Macki-Strauss: Introduction to Optimal Control Theory.

Malitz: Introduction to Mathematical Logic.

Marsden/Weinstein: Calculus I, II, III. Second edition.

Martin: Counting: The Art of Enumerative Combinatorics.

Martin: The Foundations of Geometry and the Non-Euclidean Plane.

Martin: Geometric Constructions.

Martin: Transformation Geometry: An Introduction to Symmetry.

Millman/Parker: Geometry: A Metric Approach with Models. Second edition.

Moschovakis: Notes on Set Theory.

Owen: A First Course in the Mathematical Foundations of Thermodynamics.

Palka: An Introduction to Complex Function Theory.

Pedrick: A First Course in Analysis.

Peressini/Sullivan/Uhl: The Mathematics of Nonlinear Programming.

Undergraduate Texts in Mathematics

9781461265450